CMOS CURRENT AMPLIFIERS

THE KLUWER INTERNATIONAL SERIES IN ENGINEERING AND COMPUTER SCIENCE

ANALOG CIRCUITS AND SIGNAL PROCESSING
Consulting Editor: **Mohammed Ismail**. *Ohio State University*

Related Titles:

DESIGN OF LOW-VOLTAGE LOW-POWER CMOS ΔΣA/D CONVERTERS, *Vincenzo Peluso, Michiel Steyaert, Willy Sansen*: ISBN: 0-7923-8417-2

THE DESIGN OF LOW-VOLTAGE, LOW-POWER SIGMA-DELTA MODULATORS, *Shahriar Rabii, Bruce A. Wooley*; ISBN: 0-7923-8361-3

TOP-DOWN DESIGN OF HIGH-PERFORMANCE SIGMA-DELTA MODULATORS, *Fernando Medeiro, Angel Pérez-Verdú, Angel Rodríguez-Vázquez*; ISBN: 0-7923-8352-4

DYNAMIC TRANSLINEAR AND LOG-DOMAIN CIRCUITS: *Analysis and Synthesis*, *Jan Mulder, Wouter A. Serdijn, Albert C. van der Woerd, Arthur H. M. van Roermund*; ISBN: 0-7923-8355-9

DISTORTION ANALYSIS OF ANALOG INTEGRATED CIRCUITS, *Piet Wambacq, Willy Sansen*; ISBN: 0-7923-8186-6

NEUROMORPHIC SYSTEMS ENGINEERING: *Neural Networks in Silicon*, edited by *Tor Sverre Lande*; ISBN: 0-7923-8158-0

DESIGN OF MODULATORS FOR OVERSAMPLED CONVERTERS, *Feng Wang, Ramesh Harjani*, ISBN: 0-7923-8063-0

SYMBOLIC ANALYSIS IN ANALOG INTEGRATED CIRCUIT DESIGN, *Henrik Floberg*, ISBN: 0-7923-9969-2

SWITCHED-CURRENT DESIGN AND IMPLEMENTATION OF OVERSAMPLING A/D CONVERTERS, *Nianxiong Tan*, ISBN: 0-7923-9963-3

CMOS WIRELESS TRANSCEIVER DESIGN, *Jan Crols, Michiel Steyaert*, ISBN: 0-7923-9960-9

DESIGN OF LOW-VOLTAGE, LOW-POWER OPERATIONAL AMPLIFIER CELLS, *Ron Hogervorst, Johan H. Huijsing*, ISBN: 0-7923-9781-9

VLSI-COMPATIBLE IMPLEMENTATIONS FOR ARTIFICIAL NEURAL NETWORKS, *Sied Mehdi Fakhraie, Kenneth Carless Smith*, ISBN: 0-7923-9825-4

CHARACTERIZATION METHODS FOR SUBMICRON MOSFETs, edited by *Hisham Haddara*, ISBN: 0-7923-9695-2

LOW-VOLTAGE LOW-POWER ANALOG INTEGRATED CIRCUITS, edited by *Wouter Serdijn*, ISBN: 0-7923-9608-1

INTEGRATED VIDEO-FREQUENCY CONTINUOUS-TIME FILTERS: *High-Performance Realizations in BiCMOS*, *Scott D. Willingham, Ken Martin*, ISBN: 0-7923-9595-6

FEED-FORWARD NEURAL NETWORKS: *Vector Decomposition Analysis, Modelling and Analog Implementation*, *Anne-Johan Annema*, ISBN: 0-7923-9567-0

FREQUENCY COMPENSATION TECHNIQUES LOW-POWER OPERATIONAL AMPLIFIERS, *Ruud Easchauzier, Johan Huijsing*, ISBN: 0-7923-9565-4

ANALOG SIGNAL GENERATION FOR BIST OF MIXED-SIGNAL INTEGRATED CIRCUITS, *Gordon W. Roberts, Albert K. Lu*, ISBN: 0-7923-9564-6

INTEGRATED FIBER-OPTIC RECEIVERS, *Aaron Buchwald, Kenneth W. Martin*, ISBN: 0-7923-9549-2

MODELING WITH AN ANALOG HARDWARE DESCRIPTION LANGUAGE, *H. Alan Mantooth, Mike Fiegenbaum*, ISBN: 0-7923-9516-6

LOW-VOLTAGE CMOS OPERATIONAL AMPLIFIERS: *Theory, Design and Implementation*, *Satoshi Sakurai, Mohammed Ismail*, ISBN: 0-7923-9507-7

ANALYSIS AND SYNTHESIS OF MOS TRANSLINEAR CIRCUITS, *Remco J. Wiegerink*, ISBN: 0-7923-9390-2

COMPUTER-AIDED DESIGN OF ANALOG CIRCUITS AND SYSTEMS, *L. Richard Carley, Ronald S. Gyurcsik*, ISBN: 0-7923-9351-1

HIGH-PERFORMANCE CMOS CONTINUOUS-TIME FILTERS, *José Silva-Martínez, Michiel Steyaert, Willy Sansen*, ISBN: 0-7923-9339-2

CMOS CURRENT AMPLIFIERS

by

Giuseppe Palmisano
Gaetano Palumbo
Salvatore Pennisi

Faculty of Engineering
University of Catania, Italy

KLUWER ACADEMIC PUBLISHERS
Boston / Dordrecht / London

Distributors for North, Central and South America:
Kluwer Academic Publishers
101 Philip Drive
Assinippi Park
Norwell, Massachusetts 02061 USA
Telephone (781) 871-6600
Fax (781) 871-6528
E-Mail <kluwer@wkap.com>

Distributors for all other countries:
Kluwer Academic Publishers Group
Distribution Centre
Post Office Box 322
3300 AH Dordrecht, THE NETHERLANDS
Telephone 31 78 6392 392
Fax 31 78 6546 474
E-Mail <orderdept@wkap.nl>

 Electronic Services <http://www.wkap.nl>

Library of Congress Cataloging-in-Publication Data

A C.I.P. Catalogue record for this book is available
from the Library of Congress.

Copyright © 1999 by Kluwer Academic Publishers

All rights reserved. No part of this publication may be reproduced, stored in a retrieval system or transmitted in any form or by any means, mechanical, photo-copying, recording, or otherwise, without the prior written permission of the publisher, Kluwer Academic Publishers, 101 Philip Drive, Assinippi Park, Norwell, Massachusetts 02061

Printed on acid-free paper.

Printed in the United States of America

To our wives,

        Alia

        Michela

        Stefania

Contents

Acknowledgments	ix
Abbreviations	xi
Preface	xiii

1. OPERATIONAL AMPLIFIERS — 1
- 1.1 THE OPERATIONAL AMPLIFIER SET — 2
- 1.2 OP-AMP CONFIGURATIONS — 9
- 1.3 THE GAIN BANDWIDTH TRADE-OFF — 14
- 1.4 CURRENT AMPLIFIERS USING VOLTAGE OP-AMPS — 17
- 1.5 SIGNAL PROCESSING WITH CURRENT OP-AMPS — 21
- 1.6 TOWARDS A TRUE COA — 23
- 1.7 PERFORMANCE PARAMETERS OF CURRENT AMPLIFIERS — 26
  - 1.7.1 Further Comments — 33
- REFERENCES — 40

2. LOW-DRIVE CURRENT AMPLIFIERS — 45
- 2.1 INPUT STAGES — 48
  - 2.1.1 The CCII — 48
  - 2.1.2 Class A Input Stages — 51
  - 2.1.3 Class AB Input Stages — 63
- 2.2 CLASS 'A' CURRENT OUTPUT STAGES — 69
  - 2.2.1 Output Stages for COAs — 70
  - 2.2.2 Output Stages for VFCOAs — 74
- 2.3 DESIGN EXAMPLES — 74
  - 2.3.1 COA Configurations — 75
  - 2.3.2 VFCOA Configurations — 79
- 2.4 CURRENT COMPARATORS — 84
  - 2.4.1 High-Speed Approaches — 85
  - 2.4.2 Design Control Considerations — 88
  - 2.4.3 Offset Compensation — 89
  - 2.4.4 Design Examples — 97
- REFERENCES — 102

3. HIGH-DRIVE CURENT AMPLIFIERS — 107
- 3.1 CLASS AB CURRENT OUTPUT STAGES — 108
  - 3.1.1 Configurations — 111
- 3.2 HARMONIC DISTORTION DUE TO CHANNEL-LENGTH MODULATION — 113
  - 3.2.1 COS Based On Regular Cascoded Mirrors — 113
  - 3.2.2 COS Based On Cascoded Mirrors With Dynamic Matching — 115

	3.2.3 COS Based On Cascoded Mirrors With Improved Dynamic Matching	116
	3.2.4 COS based on Active-Gain Enhanced Cascoded Mirrors	118
	3.2.5 Simulation Results	119
3.3	HARMONIC DISTORTION DUE TO MISMATCHES	121
	3.3.1 Threshold Voltage Mismatches	121
	3.3.2 Transconductance Parameter Mismatches	124
3.4	DESIGN EXAMPLES	125
	3.4.1 A VFCOA Configuration	125
	3.4.2 COA Configurations	136
3.5	A VERSATILE FULLY DIFFERENTIAL COA	146
3.6	MEASUREMENT STRATEGIES	150

APPENDIX 3.A Harmonic Distortion in class AB COSs 153
APPENDIX 3.B Accurate Determination of HD_2 155
REFERENCES 157

Index 159

ACKNOWLEDGEMENTS

CMOS Current Amplifiers was originally the title of a Doctoral thesis by S. Pennisi with G. Palmisano as supervisor. The thesis dealt mainly with high-drive CMOS current amplifiers for off-chip loads.

This book is an extended version of the thesis. It includes new topics such as Current Comparators and Conveyors, and several published works of the authors in the field of current-mode techniques. In addition, state-of-the-art achievements have been detailed by carefully considering significant contributions from other researchers. Therefore, we would like to thank the many authors who with their original ideas have produced many of the meaningful results in this field.

We would like to acknowledge the CNR (Italian National Research Council) and MURST (Ministero dell'Università e della Ricerca Scientifica e Tecnologica) for providing part of our funding.

We are grateful to EUROPRACTICE and AMS for software CAD support and technology facilities.

A special thanks is due to Prof. Bruun, for our useful exchange of information with him.

Finally, we would like to thank our friend Dr. Papalia, for his precious contribution in revising the style and syntax of the text.

Giuseppe Palmisano
Gaetano Palumbo
Salvatore Pennisi

ABBREVIATIONS

A/D	Analog-to-Digital Converter
BJT	Bipolar Junction Transistor
CCCS	Current-Controlled Current-Source
CCII	Second Generation Current Conveyor
CCVS	Current-Controlled Voltage-Source
CFOA	Current-Feedback Voltage Operational Amplifier
CMRR	Common-Mode Rejection Ratio
COA	Current Operational Amplifier
COS	Current Output Stage
D/A	Digital-to-Analog Converter
DUT	Device Under Test
GBW	Gain-Bandwidth Product
IC	Integrated Circuit
MOS	Metal Oxide Semiconductor
Op-Amp	Operational Amplifier
OTA	Operational Transconductance Amplifier
PSRR	Power Supply Rejection Ratio
SR	Slew Rate
TCOA	Transconductance Operational Amplifier
THD	Total Harmonic Distortion
TROA	Transresistance Operational Amplifier
VCCS	Voltage-Controlled Current-Source
VCVS	Voltage-Controlled Voltage-Source
VFCOA	Voltage-Feedback Current Operational Amplifier
VOA	Voltage Operational Amplifier

PREFACE

Over the last few years current-mode signal processing has been extensively investigated. Works have been published demonstrating that state-of-the-art current-mode analog design can provide solutions to many circuit and system problems. In this field, current mirrors have been the natural choice for use as elementary building blocks, since they allow simple and elegant open-loop circuits to be designed with low cost processes. On the other hand, the approaching maturity of current-mode techniques and increasing demands on performance have lead to high-gain current-input current-output circuits for use in accurate closed-loop configurations. As can be expected from amplifiers exploiting current-mode techniques, performance in terms of low voltage, slew rate and bandwidth can in principle be maximized.

Furthermore, there is also need for power stages capable of delivering high bipolar currents (several milliamperes) into off-chip loads. To this end, the high-drive current amplifier becomes the natural front-end block for current-mode IC's.

The book presents design strategies for high performance current amplifiers based on the CMOS technology, preferred for VLSI analog processing. After an introduction on various architectures of operational amplifiers, the operating principles of the current amplifier are outlined. The main focus will be to provide the reader with simple and compact design equations for use in a pencil and paper design and the following simulation step. More complex analytical derivations will be introduced only for the evaluation of fundamental non-idealities due to second order effects. Therefore, the length of the book is self-contained and presentation is kept to a reasonably accessible level. The book was written to be used by postgraduate students who are already familiar with voltage operational amplifiers and who want to extend their knowledge to current amplifiers. Moreover, it is a valid reference for analog IC designers who desire to implement this unconventional amplifier and to better comprehend its practical applications and limitations. The outline of the text is as follows:

Chapter 1 introduces the general aspects of Current Amplifiers. After a preliminary classification of operational amplifiers, ideal blocks and models are discussed for different architectures and a first high-level comparison between traditional amplifiers and current amplifiers is made. Analysis and examples of basic circuits as well as signal processing applications involving current amplifiers are also given. Non-idealities and second-order effects causing limitations in performance are then discussed and evaluated.

Chapter 2 focuses on low-drive Current Amplifiers. Several design examples for Current Conveyors and class A Current Amplifiers are discussed in detail and design equations are presented for the main performance parameters which allow a good trade-off between requirements. Moreover, high-performance solutions for high bandwidth and low voltage capability are also considered. Finally, Current Comparators with progressively enhanced performance are reported and analyzed critically.

Chapter 3 deals with current amplifiers for off-chip loads. Several class AB current-mode output stages are discussed and design strategies which improve performance are presented. A detailed analysis of non-ideal effects is carried out with particular emphasis on linearity. Design examples are given and circuit arrangements for further developments are included. Experimental measurements are finally reported.

Chapter 1

OPERATIONAL AMPLIFIERS

Amplifiers with high open-loop gain, most commonly termed Operational Amplifiers (op-amps) [1]-[4], are unquestionably the most useful and flexible building blocks in analog signal processing. Indeed, the use of these devices in a negative feedback loop allows a large variety of transfer functions to be simply implemented, and whose properties and accuracy are essentially independent of the large but inaccurate open-loop gain of the op-amp.

Traditionally, signal processing has been restricted to voltage-mode operations only. As a result, the voltage op-amp (VOA) has been the dominant architecture used by analog designers and the most commercially available from IC manufacturers.

However, in recent years the so called current-mode approach [5]-[8] has attracted ever more attention. Circuits are classified as current-mode if the information medium is represented by time-varying currents. This approach is particularly useful in the IC environment which is predominantly capacitive. Under this condition, speed is maximized by driving currents rather than by voltages. Moreover, low-voltage operations can be associated with current-mode circuits since voltage swings are minimized, while the signal range depends on the impedance level chosen by the designer and is no longer directly restricted by the supply voltage.

In effect, the last decade has been characterized by an increasing number of current-mode applications both in the digital and analog area. For example, current-mode has been profitably employed in high-speed circuits including current-mode logic [9]-[16] current-mode D/A and A/D converters [17]-[27] as well as in linear [28]-[40] and log-domain [41]-[51] filter applications, and in translinear circuits [52]-[58].

In this context, amplifier architectures with current-input current-output capability represent useful building blocks that can play the same role as VOAs do in traditional voltage-mode applications.

In this chapter, after a brief introduction on ideal amplifiers, their models and basic theory, we will review practical amplifier architectures and their high-level behavior. The voltage op-amp, transconductance op-amp and current-feedback op-amp will be discussed together with their favorable features and principal limitations. Then, reported techniques to provide conventional voltage amplifiers with current capability will be reviewed and applications of current amplifiers to signal processing will be also treated. Finally, the internal architecture of true current-mode amplifiers and their main performance parameters will be introduced.

1.1 THE OPERATIONAL AMPLIFIER SET

The concept of an ideal operational amplifier was first introduced by Tellegen in 1954 [59]. According to this definition an ideal op-amp is a two-port device with four associated variables, V_i-I_i and V_o-I_o at the input and output port, respectively, which exhibits an infinite power gain between the two ports. This means the input variables are both zero whereas the output ones are arbitrary. The input port is consequently said to exhibit a virtual short-circuit. Indeed, the input terminals have the same voltage and no current flows through them. The block diagram of an ideal op-amp is shown in Fig. 1.1.

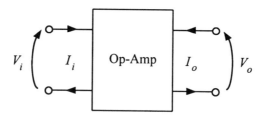

Fig. 1.1. Block diagram of an ideal op-amp

The transfer properties of this device only become well-defined if an external network allows feedback from the output to the input port, as depicted in Fig. 1.2.

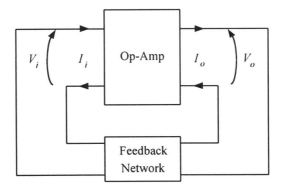

Fig. 1.2. Amplifier with feedback network

In practice, an op-amp belongs to one of the four types listed below:

- voltage op-amp (VOA): a voltage controlled voltage source (VCVS) with both infinite voltage gain and input resistance, and zero output resistance
- current op-amp (COA): a current controlled current source (CCCS) with both infinite current gain and output resistance, and zero input resistance
- transresistance op-amp (TROA): a current controlled voltage source (CCVS) with infinite transresistance gain, and both zero input and output resistances
- transconductance op-amp (TCOA): a voltage controlled current source (VCCS) with either infinite transconductance gain, input and output resistances

Of course, ideal performance in terms of bandwidth, linearity, input and output swing, etc., are also requirements for any kind of ideal op-amp.

The block diagram and the small-signal equivalent circuit for each of these devices are reported in Fig. 1.3. Input (r_i) and output (r_o) resistances are also included. Table 1.1 summarizes the characteristics of the four ideal op-amps which exhibit an infinite power gain between the input and output ports.

It is self-evident that the four devices can be arranged into two dual pairs. The VOA and the COA form one pair, while TROA and TCOA constitute the other pair (this is the particular case of application of the *adjoint networks* theorem, briefly described in section 1.5). Indeed, from Table 1.1, it is clear that the former pair has an input resistance which is reciprocal to the output resistance and that the gain is the ratio of the same kind of variable, while the second pair has equal input and output resistances and the gain is a ratio of different variables.

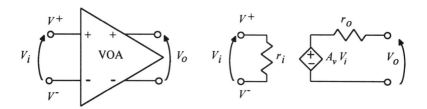

Fig. 1.3a. Voltage op-amp symbol and equivalent circuit

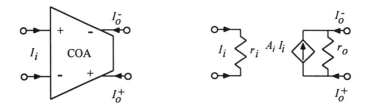

Fig. 1.3b. Current op-amp symbol and equivalent circuit

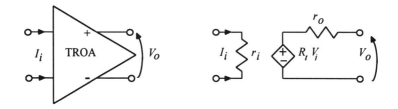

Fig. 1.3c. Transresistance op-amp symbol and equivalent circuit

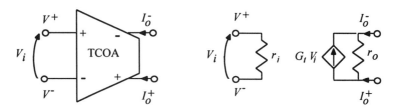

Fig. 1.3d. Transconductance op-amp symbol and equivalent circuit

Table 1.1.
Ideal op-amps and their characteristics

Op-amp type	r_i	r_o	Open-loop gain
VOA	∞	0	$A_v = V_o/V_i$
COA	0	∞	$A_i = I_o/I_i$
TROA	0	0	$R_t = V_o/I_i$
TCOA	∞	∞	$G_t = I_o/V_i$

Amplification and signal conversion (i.e., voltage to current and current to voltage conversion) are the most frequently encountered functions in analog signal processing. These functions can be implemented with a feedback op-amp by a suitable choice of the external feedback network.

There are four possible types of closed-loop amplifiers which differ in the combinations of input source and output drive

- voltage to voltage (V-V) amplifier
- current to current (I-I) amplifier
- voltage to current (V-I) amplifier
- current to voltage (I-V) amplifier

The closed-loop configurations for each kind of amplifier are illustrated in Fig. 1.4, where the symbol of the ideal op-amp was used.

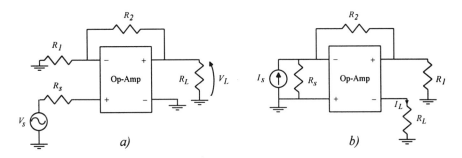

***Fig. 1.4.** Feedback amplifiers: a) V-V ; b) I-I*

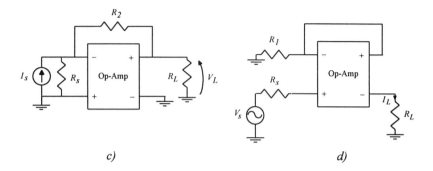

Fig. 1.4. *Feedback amplifiers: c) I-V ; d) V-I*

Table 1.2 summarizes the transfer functions which result for each kind of feedback amplifier.

Table 1.2.
Transfer functions of amplifiers in Fig. 1.4

V-V	I-I	I-V	V-I
$1+R_2/R_1$	$1+R_2/R_1$	R_2	$1/R_1$

It can be observed that the above transfer functions only depend on the values of R_1 and/or R_2 regardless of the source and load resistances. This is a desirable feature which closely approximates the performance of an ideal amplifier, since it reduces interaction between cascaded active circuits and improves control over the loop gain frequency response (module and phase). This feature greatly simplifies design from the system to circuit point of view.

At this point one may be tempted to conclude that any of the four op-amps might alternatively be used to implement the four types of feedback amplifiers. However, this is not the case if we consider non-ideal op-amps with finite (albeit large) open-loop gain, even with ideal internal resistances. In fact, under these assumptions, most of the 16 closed-loop configurations, obtained by replacing the ideal op-amp with one of the four specified op-amps in Fig. 1.3, will exhibit a loop gain which is dependent on the source and/or load resistances.

Since the closed-loop gain and bandwidth are strictly related to the loop gain (this important aspect of closed-loop systems will again be briefly reviewed in section 1.3), they will also depend on the source and/or load resistances. More specifically, this detrimental condition characterizes all the configurations in which the op-amp is a different type to the feedback amplifier [60], [61].

We will give two examples to illustrate this concept. Firstly, we will analyze the loop gain of a I-I feedback amplifier implemented using a VOA and then the loop gain of the same I-I amplifier made using a COA.

Let us consider the circuit in Fig. 1.5a where the ideal op-amp of Fig. 1.4b is replaced by a VOA.

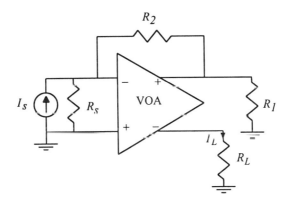

Fig. 1.5a. VOA employed to realize a I-I amplifier

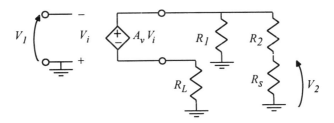

Fig. 1.5b. Equivalent circuit for the loop-gain computation of a I-I amplifier realized with a VOA

We consider the equivalent circuit in Fig 1.5b (obtained by breaking the loop at the inverting input of the VOA) to evaluate the loop gain, T, which is given by

$$T = -\frac{V_2}{V_1} = A_v \frac{R_1 R_s}{(R_1 + R_L)(R_2 + R_s) + R_1 R_L} \quad (1.1)$$

The reader can verify this result directly by inspection. The loop gain clearly depends on both source and load resistances.

Let us now consider the implementation of the same I-I amplifier in a more "natural" mode, i.e. by using a COA. In this case the loop gain can be computed by considering the equivalent circuit in Fig. 1.6.

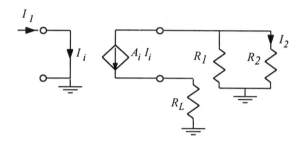

Fig. 1.6. *Equivalent circuit for the loop-gain computation of a I-I amplifier realized with a COA.*

In this case the loop-gain is totally free from source or load effects, as (1.2) shows

$$T = -\frac{I_2}{I_1} = A_i \frac{R_1}{R_1 + R_2} \qquad (1.2)$$

To preserve simplicity, the above analysis has been restricted to op-amps with finite gain but with ideal input-output resistances. However, it can also be demonstrated that the considerations developed above remain valid with real input-output resistances as well.

To conclude, we can say that the best performance is obtained by following "natural laws" and that the use of a VOA is not the prime choice in implementing current-mode transfer functions. This perhaps represents the principal motivation that lead researchers to design more appropriate op-amp architectures which could be profitably used in current-mode signal processing.

1.2 OP-AMP CONFIGURATIONS

In the previous section we dealt with the four types of ideal op-amps which satisfy the general definition. Moreover, we also showed that, in real cases, each one performs better if used in its natural way (i.e., a VOA to implement a V-V amplifier, a TROA to implement a I-V amplifier, etc.). Nevertheless, given the wide diffusion of voltage signal processing, the VOA has until recently been the dominant architecture for analog designers and the one most commonly available from IC manufacturers. This ubiquitous device mainly bases its popularity on the number of attractive features which it allows. In fact, the differential pair input stage is well suited to rejecting common-mode signals. Moreover, a VOA only requires a single-ended output to provide negative feedback and output drive capability at the same time as shown in Figs. 1.4a and 1.4c. In these figures one of the two output terminals is grounded in both voltage-output applications. Of course, implementing a fully differential output stage is more difficult than designing a single-ended output stage. Figure 1.7 shows the internal architecture of a typical single-ended VOA. It is made up of a transconductance input stage (usually implemented with a differential pair), a second stage to enhance the overall amplifier gain and an output stage (usually implemented with a Class AB voltage follower) to isolate the op-amp output from the load.

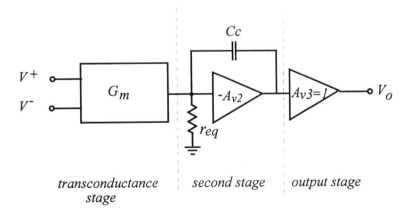

Fig.1.7. Typical architecture of a single-ended VOA

Resistance r_{eq} is the equivalent output resistance of the input stage and C_C is the compensation capacitor. Generally, Miller compensation is adopted because it leads to a small capacitance and provides the well-known pole splitting [62], [63].

For the circuit in Fig. 1.7 the DC open-loop gain and the pole frequency are

$$A_v = G_m r_{eq} A_{v2} \qquad (1.3)$$

$$\omega_o = \frac{1}{r_{eq} A_{v2} C_C} \qquad (1.4)$$

However, it is well known that the performance of a VOA is limited by a fixed gain-bandwidth product and by an internal slew-rate whose maximum value is determined by the ratio of a quiescent current to the compensation capacitance.

In the last two decades, emerging MOS technologies (initially NMOS and now almost exclusively CMOS) have lead to the development of transconductance operational amplifiers (commonly termed OTAs) principally for use in fully-integrated filtering. These amplifiers are integrated versions of what we referred to as TCOA in section 1.1. Since OTAs are used in voltage-mode applications, with the exception of fully differential circuits, their design can be simplified to provide a single ended output.

VOA and OTA architectures are strongly related since an OTA can be considered a VOA without the final buffer stage. Indeed, an output stage can be avoided in those applications where the load is a small capacitance requiring only low current levels. Moreover, a capacitive load leaves the DC loop gain unchanged only affecting its phase. The use of CMOS technology has developed continuously because of its ability to support high quality analog and digital circuitry, leading to the design of high-performance op-amps for off-chip loads. Due to their reduced drive capability compared to BJT technologies, CMOS power amplifiers have been even better implemented in transconductance output stages (i.e., common source output stages instead of voltage follower stages) [64]-[67]. In these applications the main goals are drive capability and power conversion efficiency. To this end, the output stage has to provide a large input and output swing but without requiring a voltage gain. Usually, it behaves like a low-gain inverting amplifier.

More recently, we have witnessed the affirmation of a novel op-amp architecture now available from several of the specialist analogue semiconductor manufactures. These op-amps are generally referred to as Current-Feedback Operational Amplifiers (CFOAs) [2], [7] and [68]-[70], and represent an evolution in the architecture of the voltage-mode op-amps, which have otherwise remained much the same over the years. Figure 1.8 shows the CFOA symbol and equivalent circuit. Like almost all current-mode circuits, a CFOA can be conveniently described in terms of the well-known Second Generation Current Conveyor (CCII) which essentially consists of two coupled voltage and current unitary buffers (readers who are not familiar with this building block, can refer to the next chapter for a brief introduction).

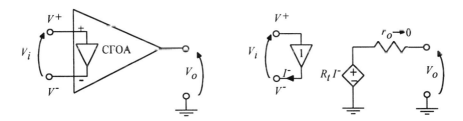

Fig 1.8. CFOA symbol and equivalent circuit

In Fig. 1.9 the typical architecture of a CFOA is illustrated. It is made up of a CCII-based input stage (that performs a voltage following action between terminal Y to X and a current following action between X and Z), and an output stage with the classical buffering function.

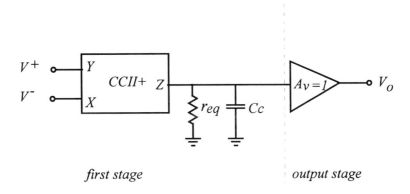

Fig.1.9. Typical architecture of a CFOA

By applying a current signal to the (low impedance) inverting input terminal, the DC open-loop transresistance gain can be found

$$R_t = r_{eq} \qquad (1.5)$$

The dominant-pole frequency is

$$\omega_o = \frac{1}{r_{eq}C_C} \qquad (1.6)$$

Implementations of high-performance CFOAs have become possible thanks to the availability of high quality complementary bipolar transistors provided by advanced BJT processes. Here, the traditional differential pair input stage has been abandoned for a complementary common emitter/common base stage. This non-conventional input stage allows a nominally unlimited slew rate to be achieved thus yielding a large signal response which is superior to that of most voltage op-amps. Moreover, this device provides the well-known constant bandwidth versus variable closed-loop gain property.

These features have led to conventional voltage op-amps being replaced with CFOAs in some high speed applications. However, these op-amps are affected by some drawbacks. First, the input voltage follower works outside the feedback loop and can be a heavy source of distortion especially in unity-gain configurations with high signal levels. Secondly, CFOAs exhibit high input offsets and low common-mode rejection. Partially because of these limitations and given the inherent low transconductance of MOS transistors, only few CFOAs have been presented up to now in CMOS technology.

We can simply illustrate the constant-bandwidth property offered by a CFOA by considering the closed-loop configuration shown in Fig. 1.10. The loop gain can be found by shorting the input source and breaking the loop at the inverting input. The equivalent circuit obtained by updating the equivalent resistance levels is shown in Fig. 1.11 (where ideal input and output CFOA resistances are considered).

The loop gain is given by

$$T_o = -\frac{I_2}{I_1} = \frac{R_t}{R_2} \qquad (1.7)$$

which shows that the loop gain depends on the transresistance gain, R_t, and R_2. Since the closed-loop bandwidth is proportional to the loop gain, and the closed-loop gain can be set by changing only R_1, a constant bandwidth behavior is achieved.

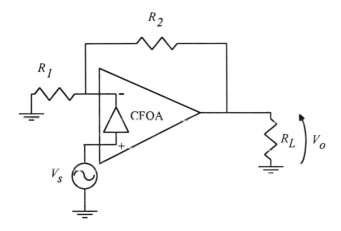

Fig.1.10. *Feedback configuration of a CFOA*

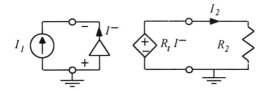

Fig. 1.11. *Equivalent circuit for the loop-gain computation of the circuit in Fig. 1.10*

1.3 THE GAIN BANDWIDTH TRADE-OFF

The constant bandwidth property exhibited by the CFOA can easily be explained and generalized using the traditional block diagram for a feedback system shown in Fig. 1.12. We have chosen not use this representation in the future because it is difficult to take into account the loading effects of the feedback network on the basic amplifier. However, this approach is quite suitable from a high-level point of view and helps to get a simple focus on the argument [61].

$A(s)$ represents the open-loop gain of the amplifier, α is the input attenuation and β the attenuation introduced by the feedback network. $A(s)$ can be one of those listed in the last column of Table 1.1, depending on the type of op-amp chosen.

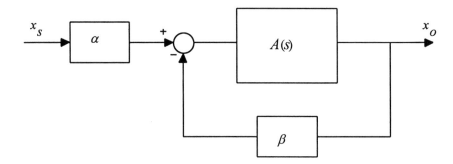

Fig.1.12. Block diagram of a feedback system

From this diagram we find the expressions of the loop gain, $T(s)$, and the closed-loop gain, $A_{cl}(s)$

$$T(s) = \beta A(s) \tag{1.8}$$

$$A_{cl}(s) = \frac{x_o}{x_s} = \frac{\alpha A(s)}{1 + T(s)} \tag{1.9}$$

It can be observed that the loop gain is dimensionless in all possible cases. Moreover, if $T(0)$ is much greater than 1, the DC closed-loop gain can be approximated to

$$A_{cl}(0) \cong \frac{\alpha}{\beta} \tag{1.10}$$

Assuming that the op-amp is characterized by a single dominant pole its open-loop gain can be expressed as

$$A(s) = A_o \frac{1}{1 + s/\omega_o} \qquad (1.11)$$

where ω_o is the angular frequency of the dominant pole and A_o is the DC gain.

Substituting (1.11) in (1.9) we obtain the expressions for the closed-loop amplifier bandwidth, ω_{cl}, and gain-bandwidth product, GBW,

$$\omega_{cl} = \omega_o [1 + T(0)] \qquad (1.12)$$

$$GBW = \frac{|\alpha| A_o}{1 + T(0)} \omega_o [1 + T(0)] = |\alpha| A_o \omega_o \qquad (1.13)$$

The above equations hold for any feedback system. Provided that parameter α is constant, (1.13) asserts the well-known property of GBW to be independent of the closed-loop gain (i.e., from β). Note that GBW has the frequency dimension for the VOA and COA only. For the TROA it has the dimension of the reciprocal of a capacitance, and for the TCOA it has the dimension of the reciprocal of an inductance.

Furthermore, combining (1.8) and (1.12) we see that the closed-loop bandwidth is approximately given by $\beta A_o \omega_o$. Therefore, by leaving parameter β (and hence the closed-loop bandwidth) constant, one can change the closed-loop gain simply by changing parameter α, as according to (1.10). This is exactly what happens in a CFOA. On the contrary, in some applications with VOAs (e.g., a non-inverting configuration where α is unitary) the closed-loop gain is set by changing β and, hence, the loop gain and the closed-loop bandwidth.

In [60] it was shown that all the four types of op-amps used in their natural mode (i.e. to implement the same type of feedback amplifier) are characterized by α equal to 1 and consequently, exhibit constant gain-bandwidth properties. For example, we can return to (1.2), expressing the loop-gain of the I-I feedback amplifier performed by a COA. We can rewrite (1.2) in terms of the closed-loop gain A_{cl}

$$T = A_i \frac{R_1}{R_1 + R_2} = \frac{A_i}{A_{cl}} \tag{1.14}$$

By comparing (1.8) and (1.14) we get $\beta = 1/A_{cl}$, and from (1.10) $\alpha = 1$ results. Hence, any increase in the closed-loop gain leads to a decrease in the closed-loop bandwidth. The same results and considerations can be extended to the so-called constant gain-bandwidth product configurations.

Moreover, the same authors demonstrated that achieving constant bandwidth properties depends on the possibility to isolate the op-amp input and output from the source and load resistances, respectively, by using ideal current or voltage unity-gain buffers. For instance, the CFOA can be recognized as the result of applying a voltage buffer to a single-ended transresistance amplifier. A further example can be given if we return to (1.1) that expresses the loop-gain of a I-I feedback amplifier performed by a VOA. We can see that adopting two ideal current followers at the input and the output of the amplifier, is equivalent to setting resistance R_s to infinity and R_L to zero. Under these assumptions loop gain becomes

$$T(s) = A_v(s) \tag{1.15}$$

The above relation indicates that parameter β is equal to 1. Therefore, this particular configuration will achieve the maximum closed-loop bandwidth, i.e. the GBW, regardless of the value of closed-loop gain, source and load resistances. This is a very attractive solution which, however, is not real. Indeed, it is no trivial task to design high-performance followers, with approaching ideal input-output resistances and a bandwidth performance which is well beyond that of the final amplifier. Actually, when considering the fundamental process limitations which set the ultimate performance for both unity-gain buffers and op-amps, we see that constant gain-bandwidth product configurations provide higher loop gain than constant bandwidth configurations, as discussed in [71]. This means that, in practical cases, the constant bandwidth property is obtained at the expense of reduced loop gain.

The brief discussion given here reveals that today there is great interest in the op-amp area due to the demand for ever better performance. In addition, advances in process technologies have made possible the integration of novel architectures for use in special applications. However, for the remainder of this book we shall concentrate only on the realization of current operational amplifier architectures, i.e. op-amps with well-defined input and output current-capabilities. We shall start by giving a historical background.

1.4 CURRENT AMPLIFIERS USING VOLTAGE OP-AMPS

The need for amplifiers with a well-defined bipolar current output was recognized even when the current-mode approach was in its infancy. In the time when no fully integrated version of current amplifiers was available, the only alternative for system designers was to use a voltage amplifier. Several voltage-to-current amplifier designs employing the VOA have been reported in literature from the late 1970s to the end of the 1980s. The very first attempts sought to provide only a single current output facility using a VOA. The most common way to do this, as frequently encountered in textbooks [1], is to place the load in the feedback loop of the VOA as shown in Fig. 1.13. This circuit provides a very low input resistance (at the inverting terminal). The current flowing through R_L is set to V_s / R_1 regardless of the R_L value. Therefore, the circuit can be seen as a current generator between terminals A and B. Indeed, the equivalent resistance seen at these terminals is equal to $(r_i // R_1)(1 + A_v)$, where r_i and A_v are the VOA input resistance and voltage gain, respectively. However, in this solution the load cannot be referred to ground, despite this being a general requirement.

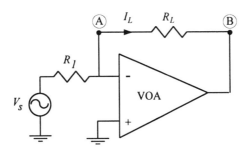

Fig. 1.13. Feedback voltage-to-current amplifier

To solve this problem Howland [1] proposed the circuit shown in Fig. 1.14.

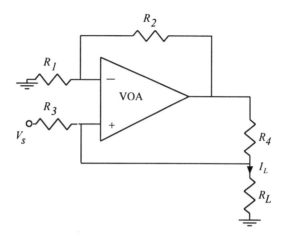

Fig. 1.14. Howland voltage-to-current converter

The current-converter will drive the grounded load with a current $I_L = V_s / R_3$ if the balancing condition $R_4 / R_3 = R_2 / R_1$ is satisfied. In addition, thanks to the combination of positive and negative feedback, the circuit exhibits a nominal infinite output resistance. Unfortunately, the positive feedback action means the circuit requires very accurate resistor matching. Indeed, even a small deviation from the ideal balancing condition could make the output resistance negative and cause instability.

To extend the output capability of conventional VOAs, the most successful technique was using *supply current sensing*, first introduced by Graeme [72] in 1974 and further refined by Rao and Haslett [73]. The approach exploits the property that the sum of the currents flowing in the VOA supply leads equals the output current, so long as the signal currents in other connections to ground are negligible. In most VOAs this condition is met and a couple of complementary current mirrors can be employed to recombine the two supply currents. The current mirrors provide a single high impedance bipolar output, as shown in Fig. 1.15. Huijsing coined the term *Operational Mirrored Amplifier* to indicate this circuit [74]. Of course, for proper current-converter operation the structure has to be configured in unity-gain closed-loop (i.e., by shorting the VOA output and inverting input). Under this arrangement the circuit can be used in a number of applications since it realizes the versatile CCII, as will be shown later.

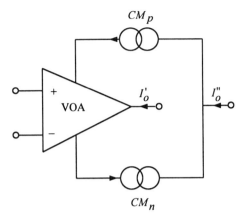

Fig. 1.15. Operational mirrored amplifier

Moreover, by including a resistive feedback network, an accurate current gain is achieved, as illustrated in Fig. 1.16. It is given by

$$\frac{I_L}{I_s} = -\left(1 + \frac{R_2}{R_1}\right) \qquad (1.16)$$

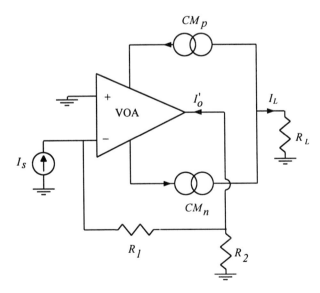

Fig. 1.16. Operational mirrored amplifier arranged to provide a current gain

The supply-sensing technique was also generalized by Toumazou et al. [6] who propose that a multi-terminal VOA be implemented leaving the drains (collectors) of the output stage externally accessible. In this manner, output-current sensing rather than whole supply-current is enabled.

Current-sensing techniques have been experimentally tested with a number of well-documented applications. Despite the use of discrete components, these circuits generally provide high bandwidth and slew rate capabilities. However, the critical parameter in these applications is linearity because it depends upon the current transfer properties of the current mirrors which are connected in open loop at the circuit output. This drawback is common to almost all class AB solutions and, due to its importance, it will be dealt with in detail in chapter 3.

We conclude this review with the simple and well-known class A current amplifier shown in Fig. 1.17. In this circuit transistor M1 and current generator I_B make up the current output stage. Apart from the sign, it is easy to find that the amplifier gain is given again by (1.16). This circuit exhibits high accuracy and linearity, but it has a very low power conversion efficiency, since the current output stage works in a class A fashion. Moreover, it could be difficult to match current I_B with the quiescent current in M1, since they are affected by different process tolerances.

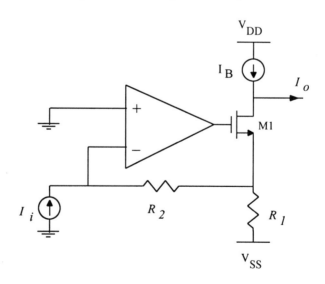

Fig. 1.17. *Simple class A current amplifier*

On the basis of what we discussed in section 1.2, the application of supply current sensing to VOAs can be recognized in the use of a current follower in the VOA output. This explains why these architectures were (sometimes surprisingly) discovered as having constant bandwidth properties. However, all these kinds of op-amp structures represent only a partial step towards the implementation of a true current-mode op-amp. In reality, the ideal COA defined in section 1.1 requires a zero open-loop input resistance and a differential current output capability, features which are not provided by the circuits discussed above.

1.5 SIGNAL PROCESSING WITH CURRENT OP-AMPS

As already mentioned, a dual relationship exists between voltage-mode and current-mode circuits. This means that almost all transfer functions implemented with active networks based on voltage op-amps can alternatively be implemented with current op-amps. Voltage-mode circuits can readily be transformed into current-mode circuits by using the principle of adjoint networks [75]-[76], that we shall briefly describe in the following.

The adjoint networks principle

Consider an arbitrary linear network consisting of passive elements (resistors, capacitors and inductors) and controlled sources. Moreover, assume that the network is driven by an independent voltage source and that the output variable is a specific voltage. By replacing each element of the network with its adjoint (or dual), the adjoint network is found, according to the following rules. The adjoint of a passive element is itself (i.e., resistors, capacitors and inductors remain unchanged). The adjoint of a voltage source is a short circuit (and viceversa). Similarly, a current source turns into a open circuit (and viceversa).

Figure 1.18 illustrates the adjoint network principle applied to the well-known non-inverting and inverting closed-loop VOA configurations.

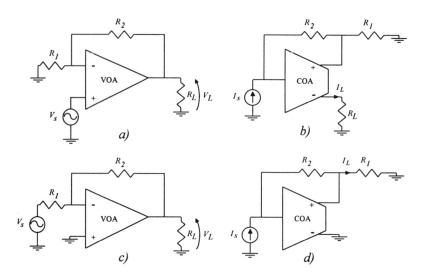

Fig.1.18. *Non-inverting, a) and b), and inverting, c) and d), feedback configurations using VOAs and COAs*

It is easy to verify that the closed-loop gain for the upper and lower circuits in Fig. 1.18 are respectively

$$A_{cl} = 1 + \frac{R_2}{R_1} \tag{1.17a}$$

$$A_{cl} = -\frac{R_2}{R_1} \tag{1.17b}$$

The current amplifiers in Fig. 1.18 possess the interesting feature which is the possibility of using non-linear resistances in the feedback network. In fact, since the voltage drop across R_1 and R_2 is the same, non-linearities are compensated for [77]. Arbel first reported this property and described a COA employing a transistorized feedback network. To achieve a gain of $n+1$ in the non-inverting configuration, resistances R_1 and R_2 are implemented with a single and a parallel of n triode-biased transistors, respectively.

The properties of an adjoint network can be inferred from the properties of the original network, without requiring any further analysis. For instance, the popular voltage-mode *RC*-active circuits can be transformed into their current-mode counterparts using the COA. The adjoint circuits have identical sensitivities and, hence, design rules and optimization strategies, developed for voltage-mode circuits, can be used directly.

An example of a current-mode integrator is given in Fig. 1.19. Two transfer functions are obtained by considering output currents I_1 or I_L

$$\frac{I_1}{I_s} = \frac{1}{sR_1C} \qquad (1.18a)$$

$$\frac{I_L}{I_s} = \frac{1+sR_1C}{sR_1C} \qquad (1.18b)$$

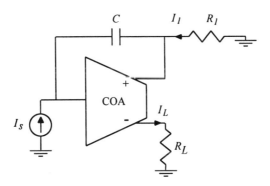

Fig. 1.19. Current-mode integrator

1.6 TOWARDS A TRUE COA

We have already introduced the ideal current op-amp in section 1.1 and recognized that, according to the adjoint network theorem, it can be represented as the dual of an ideal VOA. The theorem can be also exploited to obtain one possible COA internal architecture from that of a VOA. To this end, we noted that the design of a VOA can be considerably simplified if a single output device is considered. In fact, a single ended VOA is able to drive the load and the feedback network simultaneously. With this observation in mind it is easy to understand that a similar condition would also hold for the input port of a COA. Indeed, a COA only needs one input terminal to be connected to the input source and to the feedback network. This aspect of a current amplifier can be further exemplified by returning to Figs 1.4 b and c working as amplifiers with input current capability. It can be seen that in such applications one op-amp input terminal is always grounded, this is due to the general rule that summing or subtracting currents only requires one single node. The symbol and the circuit model of a single-input COA are reported in Fig. 1.20.

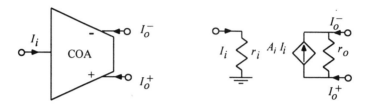

Fig 1.20. Single-input current op-amp symbol and equivalent circuit

Finally, we can apply the adjoint theorem to the VOA internal architecture, shown in Fig. 1.7, and obtain a block model of the COA as depicted in Fig. 1.21. The output voltage buffer of the VOA turns into a input current buffer. The current buffer can be also realized by using a CCII- with a grounded Y terminal. The differential input stage turns into a differential transconductance output stage. The second gain-stage can remain the same. The current gain and bandwidth of this circuit are exactly the same as the voltage gain and bandwidth of the VOA

$$A_i = A_{v2}G_m r_{eq} \tag{1.19}$$

$$\omega_o = \frac{1}{r_{eq}A_{v2}C_c} \tag{1.20}$$

where r_{eq} and C_C are the equivalent resistance at terminal Z of the CCII- and the Miller compensation capacitor, respectively. Note that the second stage, used to increase gain, is not necessary for the implementation of a current amplifier, although it allows Miller compensation to be profitably achieved. The input current buffer (possibly together with the second stage) can be also regarded as a transresistance amplifier. In fact, it accepts a low input current and provides a high output voltage.

It is curious to note that, the very first proposals of high-current-gain amplifiers [78-79] adopted differential-input devices. However, as it turned out, this was an unnecessary condition revealing how strong the voltage op-amp heritage and the voltage-mode way of thinking was.

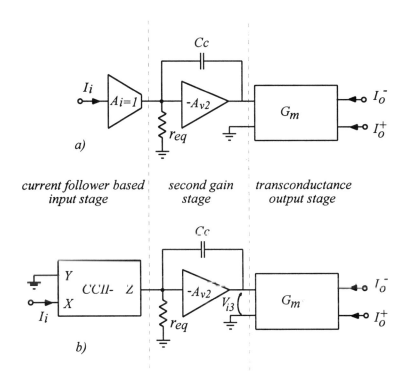

Fig. 1.21. *COA internal architecture a) with current buffer; b) with CCII*

The COA is the dual counterpart of a VOA and therefore will exhibit constant gain-bandwidth property when used in closed-loop configurations. In order to obtain a structure with constant bandwidth property, the adjoint network theorem can be applied to a CFOA as well. If we convert the equivalent circuit and block diagram of a CFOA (already depicted in Figs. 1.8 and 1.9) to current-mode we obtain a novel structure which we term the *Voltage-feedback Current Operational Amplifier* (VFCOA). Figures 1.22 and 1.23 report the symbol, equivalent circuit and block diagram of a VFCOA.

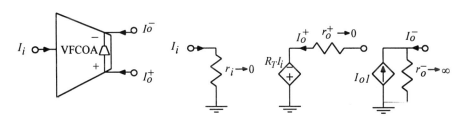

Fig 1.22. *Voltage-feedback current op-amp symbol and equivalent circuit*

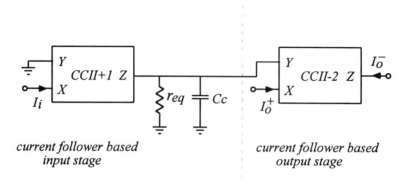

Fig. 1.23. VFCOA internal architecture

The first stage in Fig. 1.23 is the same as in a COA. The output stage is a current follower which is the dual of the input voltage follower of a CFOA. Of course, given duality the DC transresistance gain and bandwidth are the same as in a CFOA so that (1.5) and (1.6) still hold.

A final consideration is that the circuit can be regarded as the result of applying a current follower to the output of a transresistance op-amp. Consequently, the constant bandwidth versus variable gain property is explained on the basis of what was discussed in section 1.3.

1.7 PERFORMANCE PARAMETERS OF CURRENT AMPLIFIERS

In the previous sections we dealt with ideal amplifiers. This approximation is very useful in a first-order analysis and design. For example, assuming ideal open-loop gain in a real amplifier gives accurate results as long as low frequencies and low closed-loop gains are considered. However, under critical operating conditions such as high closed-loop gain, high frequency, non-ideal sources, large signal swing, etc., real values of performance parameters have to be accounted for, since they represent the final limits of operation. Therefore, we conclude the chapter by investigating the real parameters of current amplifiers which properly address the performance of this kind of circuit.

Operational Amplifiers

Differential-mode Gain

The differential gain (A_d) sets the accuracy of the closed-loop transfer function in low frequency operations. It is defined by

$$A_d = \frac{i_o^+ - i_o^-}{i_i} \qquad (1.21)$$

where $i_o^+ - i_o^-$ is the differential-mode output. Referring to the three-stage architecture in Fig. 1.21b, the differential-mode gain can be expressed as

$$A_d = r_{eq} A_{v2} \frac{i_o^+ - i_o^-}{v_{i3}} = r_{eq} A_{v2} G_{md} \qquad (1.22)$$

where G_{md} is the differential-mode transconductance gain of the output stage.

Common-mode Gain

Due to the duality with voltage amplifiers, the common-mode gain (A_c) is defined with respect to the output common mode which is $\frac{i_o^+ + i_o^-}{2}$

$$A_c = \frac{i_o^+ + i_o^-}{2i_i} \qquad (1.23)$$

Again, referring to the circuit in Fig.1.21b, A_c can be rewritten as

$$A_c = r_{eq} A_{v2} \frac{i_o^+ + i_o^-}{2v_{i3}} = r_{eq} A_{v2} G_{mc} \qquad (1.24)$$

where G_{mc} is the common-mode transconductance gain of the output stage. Output currents, i_o^+ and i_o^-, can be expressed in terms of both the differential and common-mode gains

$$i_o^+ = \left(\frac{1}{2}A_d + A_c\right)i_i \qquad (1.25)$$

$$i_o^- = \left(-\frac{1}{2}A_d + A_c\right)i_i \qquad (1.26)$$

If the common-mode gain is zero, the output currents are equal in module and with opposite direction flow.

Common-mode Rejection Ratio

As expected from the duality principle, the common-mode rejection ratio (*CMRR*) depends only on the performance of the output stage (the input stage in a voltage amplifier). It expresses the ability of the output stage to reject common-mode signals coming from its input (when it is operated in a fully-differential mode) or spurious signals from the supply lines and the substrate. *CMRR* is defined as

$$CMRR = \frac{A_d}{A_c} \qquad (1.27)$$

For the typical architecture of COA in Fig. 1.21b, it is obtained by combining (1.22) and (1.24) which yields

$$CMRR = \frac{G_{md}}{G_{mc}} \qquad (1.28)$$

Input Resistance

The input resistance (r_i, in Fig. 1.24) should ideally be zero since the circuit has to accept current signals. In most amplifiers, the input resistance is set by the transistor transconductance which, in a CMOS design, is not sufficiently low to be neglected. The input resistance is responsible for a further pole in the loop-gain transfer function which can greatly affect either stability or the closed-loop bandwidth.

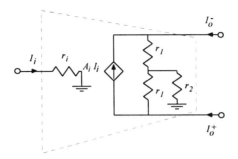

***Fig. 1.24.** Current amplifier with output resistances*

Differential and Common-mode Output Resistances

The differential (r_{od}) and common-mode (r_{oc}) output resistances can be defined by referring to Fig. 1.24. It is easy to find that

$$r_{od} = 2r_1 \qquad (1.29)$$

$$r_{oc} = \frac{r_1}{2} + r_2 \qquad (1.30)$$

To approximate an ideal current generator both the differential resistance and the common-mode resistance have to be high, as shall be explained in more detail later.

Offset Currents

There are two offset current components, i.e. the differential and the common-mode offset currents. Referring to Fig. 1.25 where the amplifier outputs are short-circuited to ground, the differential offset, I_{OSd}, is the input current needed to set the differential output current ($I_o^+ - I_o^-$) to zero and the common-mode offset, I_{OSc}, is the output current which sets to zero the common mode output current ($\frac{I_o^+ + I_o^-}{2}$).

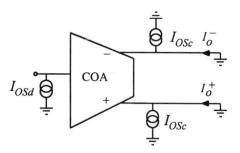

Fig. 1.25. COA with offset currents

Of course, the common-mode offset is an inherent parameter of the output stage. In fact, it depends on the mismatches in the output stage and, hence, cannot be compensated by any input current.

Input Offset Voltage

Ideally, the input bias voltage of a current amplifier is the reference (ground) voltage. As can be seen in Fig. 1.26, any deviation, V_{OS}, of the input bias voltage from the ideal value causes a DC current to flow through the resistor R_1.

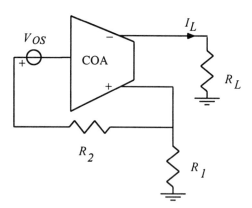

Fig. 1.26. Feedback COA with input offset voltage

Hence, the current in the load is given by

$$I_L = \frac{V_{OS}}{R_1} \qquad (1.31)$$

Of course, the offset voltage does not contribute to the output offset current if R_1 is infinitely large, as in the unity-gain buffer configuration.

Input and Output Current Ranges

The input current range (*ICR*) is the maximum input current which allows the circuit to work properly. This range is generally set by the overall transresistance gain and the voltage range at internal high impedance nodes. Like the differential input voltage range in voltage amplifiers, it is not a parameter of interest. Indeed, with high loop gain the required input range is nearly zero.

The output current range (*OCR*) is the maximum differential current which can be delivered by the output stage in short-circuit conditions. It depends on the current-drive capability of the output stage.

Slew Rate

Like in voltage amplifiers, the slew rate (*SR*) sets the maximum variation of the output current. In principle, a current amplifier with a class AB input stage (i.e., the input current buffer) behaves like a CFOA. In Fig. 1.21b, the input stage can deliver unlimited current to the internal node Z. Therefore, if a dominant-pole compensation is adopted by putting the compensation capacitor between node Z and ground, no slew-rate limitation arises. However, gain requirements with decreasing supply voltages lead to solutions with cascaded gain stages that require Miller compensation, as shown in Fig. 1.21b. In these cases, to preserve the high slew rate performance, the inverting stage (i.e., A_{V2}) should also be implemented with a class AB topology. Unfortunately, this is an expensive approach.

Power Supply Rejection Ratio

Like in conventional amplifiers, the power supply rejection ratio (*PSRR*) describes the ability of the amplifier to reject spurious signals coming from either the positive or negative power supply lines. It can be defined as

$$PSRR^{(+,-)} = \frac{\dfrac{I_o}{I_i}}{\dfrac{I_o}{(V_{dd}, V_{ss})}} \qquad (1.32)$$

and its unit is Ω.

Noise

Noise in current amplifiers can be modeled by means of three noise generators (a voltage and a current generator at the input and a current generator at the output), as illustrated in Fig. 1.27 [79]. The motivations behind this representation will be discussed in detail in sec. 1.7.1.

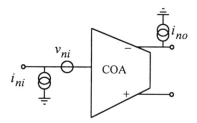

Fig. 1.27. Noise model of a COA

Here it is sufficient to say that unlike conventional VOAs, where only the input referred noise is important, in current amplifiers the noise produced by the output stage can also be significant since the output terminal connected to the load works outside the feedback loop.

We conclude this description by pointing out that the above performance parameters strictly refer only to COA. Due to its different output stage a VFCOA is characterized by a different set of parameters. Of course, differential- and common-mode gain and resistances are no longer suitable in the latter case. In addition, parameters such as offset and *PSRR* have slightly changed definitions. Aside from the above, most parameters of a COA can be extended to a VFCOA, so we shall not deal with this item any further. Instead, we shall conclude the chapter by specifying some features presented above in more detail and by providing some examples of how they affect overall closed-loop performance.

1.7.1 Further Comments

Most of the performance parameters of VOA and COA are very similar in their definition. However, due to duality some parameters in the COA such us offset current, differential- and common-mode gain and resistances belong to the output rather than the input. Noise needs also to be further detailed since it depends heavily on the specific features of the COA. In the following, we consider the properties of the output resistance and noise in typical feedback configurations. Moreover, we make final considerations regarding the constant-bandwidth property of the VFCOA.

A. Evaluation of the output resistance in a feedback amplifier

Considering the COA in a current buffer feedback configuration as shown in Fig. 1.28, the output resistance becomes

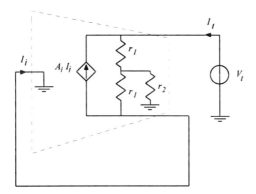

Fig. 1.28. *Equivalent circuit for the output resistance calculation in a feedback COA*

$$r_o = \frac{V_t}{I_t} = r_1 + \frac{r_1(1+2A_i)}{1+\frac{r_1}{r_2}(1+A_i)} \qquad (1.33)$$

If $r_2 \to \infty$, the common-mode output resistance in (1.30) is infinite and (1.33) simplifies to

$$r_o = 2r_1(1+A_i) \qquad (1.34)$$

that from (1.29) can be written as

$$r_o = r_{od}(1 + A_i) \tag{1.35}$$

This is the expected result for the output resistance of a shunt-series feedback amplifier: the differential output resistance is increased by the loop gain.

However, for finite values of r_2 and assuming $A_i \gg 1$, (1.33) can be approximated to

$$r_o \cong r_1 + 2r_2 = 2r_{oc} \tag{1.36}$$

This means that the common-mode output resistance sets an upper limit to the maximum achievable output resistance in a feedback COA. The same result can be obtained for the input resistance by using the popular voltage buffer implemented with a VOA.

B. Noise contributions

Let us consider the typical architecture of a COA shown in Fig. 1.29.

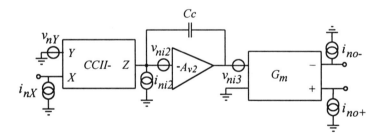

Fig. 1.29. Equivalent noise generators of the COA stages

The noise in a CCII is characterized by a voltage and a current generator (see section 2.1.1) modeled at the input by v_{nY} and i_{nX}. Voltage and current generators v_{ni2} and i_{ni2} account for the noise of the second gain stage. Voltage v_{ni2} appears at the output node (Z) of the CCII and hence its contribution to the input (X) is heavily attenuated by the high transresistance gain of the CCII. Due to the unitary current gain between X and Z, current i_{ni2} is added directly to the input stage current. Finally, we have to consider the noise contribution of the output stage. Generally, noise due to a current output stage can be divided into two components:

- the first component is referred to its input and can be represented with a voltage noise v_{ni3}. It can be further referred to the input of the input stage. This contribution to the overall noise is negligible due to the first and second stage gains.
- the second component is a noise current that adds directly to the outputs and can be modeled, in general, with two noise current generators i_{no-} and i_{no+}.

However, the non-inverting terminal is usually connected in feedback with the input. Moreover, since no current flows through the op-amp input, i_{no+} appears to the inverting output. As a result, both i_{no-} and i_{no+} can be represented at the inverting output terminal. It is interesting to note that a similar output noise component, but as a voltage generator, is also present in a VOA. However, since the output stage in a VOA is inside the feedback loop, the output noise voltage generator is heavily reduced by the loop gain.

To conclude this analysis, the noise model of a COA can be represented by three noise generators v_{ni} which is equal to v_{nY}, i_{ni} accounting for i_{nX} and i_{ni2} and i_{no} which in turn accounts for i_{no+} and i_{no-}.

C. Noise in a feedback amplifier

It is useful to evaluate overall noise power in a typical feedback amplifier based on the COA shown in Fig. 1.30. To simplify calculations, the noise related to R_1 and R_2 have been represented as current and voltage generators, respectively.

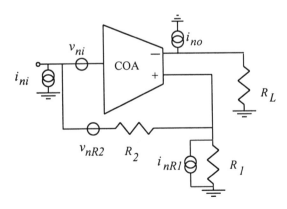

Fig. 1.30. Noise in a feedback COA

The overall output noise power, N_o, is

$$N_o = A_{cl}^2(0)\overline{i_{ni}^2} + \frac{\overline{v_{ni}^2} + \overline{v_{nR2}^2}}{R_1^2} + \overline{i_{nR1}^2} + \overline{i_{no}^2} \qquad (1.37)$$

where $A_{cl}(0)$ is the closed-loop gain.

It is interesting to compare (1.37) with the noise power of the typical VOA feedback amplifier in Fig. 1.31, assuming constant gain.

The output noise power for circuit in Fig. 1.31 is

$$N_o = A_{cl}(0)\overline{v_{ni}^2} + R_2^2\left(\overline{i_{ni}^2} + \overline{i_{nR1}^2}\right) + \overline{v_{nR2}^2} \qquad (1.38)$$

where $A_{cl}(0)$ is the non-inverting closed-loop gain.

Similarly to the noise voltage in a VOA, the contribution of the input noise current in a COA is independent of the value of R_1 and R_2. The contribution of the input noise voltage in a COA is reduced by increasing R_1 while in a VOA the noise current is increased by increasing the feedback resistance (R_2). In respect of the noise due to the resistors in the feedback network, their contribution in a COA can be reduced by increasing the resistor values. In contrast, the noise coming from the network resistors in a VOA is increased by increasing the resistor values.

Finally, if i_{no} is much greater than i_{ni} (this is the case when both the current and the aspect ratios of the output stage transistors are large), the output stage noise cannot be neglected.

In its conclusion (1.37) asserts that, with a fixed closed-loop gain, the noise performance of a current amplifier can be improved by increasing the values of the feedback resistors. This is not a surprising result since the circuit is a current processor and the noise current of a resistor is inversely proportional to its resistance. Unfortunately, this is in contrast with the need to reduce voltage biasing. Indeed, the increase in R_2 in turn increases the voltage swing at the non-inverting output and consequently a reduction in the dynamic range. Dynamic range, D, is defined as the ratio of the maximum achievable output current to the minimum usable input current. It is limited by noise and non-linearity at the lower and upper bounds, respectively.

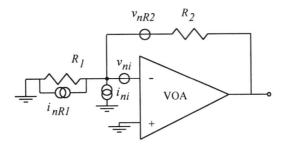

Fig. 1.31. *Noise in a feedback VOA*

D. Constant bandwidth in a VFCOA

Let us again consider the typical closed-loop VFCOA configuration in Fig. 1.32:

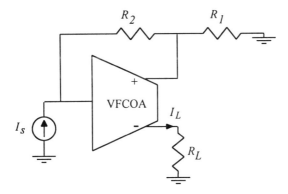

Fig. 1.32. *Closed loop VFCOA configuration*

The loop gain, $T(s)$ is given by

$$T(s) = R_t(s) \frac{R_1}{(R_2 + r_i)(R_1 + r_{o1}) + R_1 r_{o1}} \tag{1.39}$$

Equation (1.39) accounts for the input and output resistances, r_i and r_{o1}, respectively, of a real transresistance amplifier.

Assuming for $R_t(s)$ a dominant-pole behavior with pole ω_c, the closed-loop bandwidth, ω_{cl}, is approximately given by

$$\omega_{cl} \cong T(0)\omega_o \tag{1.40}$$

If the amplifier is implemented in a bipolar technology, the condition that r_i and r_{o1} are much lower than R_1 and R_2 is usually met, so that from (1.39) and (1.40) the bandwidth of the closed-loop amplifier results as:

$$\omega_{cl} \cong \frac{R_t(0)}{R_2}\omega_o \tag{1.41}$$

Equation (1.41) shows the same well-known property of a current feedback voltage amplifier (remember the duality with a voltage feedback current amplifier). The bandwidth of the closed-loop amplifier will remain independent of the closed-loop current gain provided that resistance R_2 is maintained constant.

The same result is also valid for low-drive current amplifiers in CMOS technology in which, due to the low output current, feedback resistances are set much higher than internal resistances to achieve (1.41).

In contrast, with high-drive CMOS amplifiers employing a class AB voltage follower as output stage (in the transresistance amplifier), R_1 and R_2 should be made low due to the limited output voltage swing. Therefore, for high-drive CMOS amplifiers, the condition that r_i and r_{o1} are greater than R_1 and R_2 is usually assumed. Hence, from (1.39) and (1.40), the bandwidth of the closed-loop amplifier becomes:

$$\omega_{cl} \cong R_t(0)\frac{R_1}{r_i r_{o1}}\omega_o \tag{1.42}$$

which means that the bandwidth is still constant provided that R_1 is kept constant, but the dc loop gain and ω_{cl} now depend on the input and output resistances.

It is worth noting, however, that for the unity-gain configuration in which R_1 is infinite and R_2 is zero, (1.39) becomes

$$\omega_{cl} \cong R_t(0)\frac{1}{r_i + r_{o1}}\omega_o \tag{1.43}$$

Of course, to achieve higher closed-loop gains, resistance R_2 has to be increased. However, when R_2 has the same order of magnitude of r_i the closed-loop gain asymptotically tends to a constant GBW behavior since (1.42) is no longer valid.

REFERENCES

[1] S. Franco, *Design with Operational Amplifiers and Analog Integrated Circuits*, McGraw-Hill, 1988.
[2] S. Soclof, *Design and Applications of Analog Integrated Circuits*, Prentice-Hall, 1991.
[3] J. Wait, L. Huelsman, *Introduction to Operational Amplifier Theory and Applications*, McGraw-Hill, 1992.
[4] K. Laker, W. Sansen, *Design of Analog Integrated Circuits and Systems*, Mc Graw-Hill, 1994.
[5] Z. Wang, *Current-Mode Analog Integrated Circuits and Linearization Techniques in CMOS Technology*, Hartung-Gorre Verlag Konstanz, 1990.
[6] C. Toumazou, F. Lidgey, C. Makris, "Extending Voltage-Mode OP Amps to Current-Mode Performance," *IEE Proc. Part G*, Vol.137, No.2, pp. 116-130, Apr. 1990.
[7] C. Toumazou, F. Lidgey, D. Haigh (Eds.), *Analogue IC design: the current-mode approach*, Peter Peregrinus, 1990.
[8] Z. Wang, "Current-Mode CMOS Integrated Circuits for Analog Computation and Signal Processing: A Tutorial," *Int. J. Analog Integrated Circuits and Signal Processing*, No.1, pp.287-295, 1991.
[9] G. Wilson, "Advances in Bipolar VLSI," *Proceeding of IEEE*, Vol. 78 No. 11, pp. 1707-1719, Nov. 1990.
[10] M. Kurisu, et al., "A Si Bipolar 21-GHz/320-mW Static Frequency Divider," *IEEE J. of Solid-State Circuits*, Vol. 26, No. 11, pp. 1626-1630, Nov. 1991.
[11] C. Stout, J. Doernberg, "10-Gb/s Silicon Bipolar 8:1 Multiplexer and 1:8 Demultiplexer," *IEEE J. of Solid-State Circuits*, Vol. 28, No. 3, pp. 339-343, Mar. 1993.
[12] K. Ishii, et al. "Very-High-Speed Si Bipolar Static Frequency Dividers with New T-Type Flip-Flops," *IEEE J. of Solid-State Circuits*, Vol. 30, No. 1, pp. 19-24, Jan. 1995.
[13] Z. H. Lao, et al., "A 12 Gb/s Si Bipolar 4:1-Multiplexer IC for SDH Systems," *IEEE J. of Solid-State Circuits*, Vol. 30, No. 2, pp. 129-132, Feb. 1995.
[14] L. I. Andersson, et al., "Silicon Bipolar Chipset for SONET/SDH 10 Gb/s Fiber-Optic Communication Links," *IEEE J. of Solid-State Circuits*, Vol. 30, No. 3, pp. 210-217, Mar. 1995.
[15] K. M. Sharaf, M. I. Elmasry, "Analysis and Optimization of Series-Gates CML and ECL High-Speed Bipolar Circuits," *IEEE J. of Solid-State Circuits*, Vol.31, No. 2, pp. 202-211, Feb. 1996.
[16] J. Rabaey, *Digital Integrated Circuits (A Design Perspective)*, Prentice Hall, 1996.
[17] P. Crolla,"A Fast Latching Current Comparator for 12-Bit A/D Applications," *IEEE J. of Solid-State Circuits*, Vol. SC-17, No.6, pp.1088-1093, Dec. 1982.

[18] J. Robert, P. Deval, G. Wegmann, "Novel CMOS Pipelined A/D Convertor Architecture Using Current Mirrors," *Electronics Letters*, Vol.25, No.11, pp.691-692, May 1989.

[19] D. Nairn, C. Salama, "Current-Mode Algorithmic Analog-to-Digital Converters," *IEEE J. of Solid-State Circuits*, Vol.25, No.4, pp.997-1004, Aug. 1990.

[20] Chu P. Chong, "A Technique for Improving the Accuracy and the Speed of CMOS Current-Cell DAC," *IEEE Trans. on Circuits and Systems*, Vol.37, No.10, pp.1325-1327, Oct. 1990.

[21] D. Nairn, C Salama "A Ratio-Independent Algorithmic Analog-to-Digital Converter Combining Current Mode and Dynamic Techniques" *IEEE Trans. on Circuits and Systems*, Vol.37, No.10, pp.1325-1327, Oct. 1990.

[22] Z. Wang, "Design Methodology of CMOS Algorithmic Current A/D Converters in View of Transistor Mismatches," *IEEE Trans. on Circuits and Systems*, Vol.38, No.6, pp.660-667, June 1991.

[23] Seong-Won Kim, Soo-Won Kim, "Current-Mode Cyclic ADC for Low Power and High Speed Applications," *Electronics Letters*, Vol.27, No.10, pp.818-820, May 1991.

[24] C. Wey, "Concurrent Error Detection in Current-Mode A/D Convertors," *Electronics Letters*, Vol.27, No.25, pp.2370-2372, Dec. 1991.

[25] A. Cujec, C. Salama, D. Nairn, "An Optimized Bit Cell Design for a Pipelined Current-Mode Algorithmic A/D Converter," *Int. J. Analog Integrated Circuits and Signal Processing*, No.3, pp.137-141, 1993.

[26] C. Wey, S. Krishnan, S. Sahli, "Design of Concurrent Error Detectable Current-Mode A/D Converters for Real-Time Applications," *Int. J. Analog Integrated Circuits and Signal Processing*, No.4, pp.65-74, 1993.

[27] K. Wong, K. Chao "Current-Mode Cyclic A/D Conversion Technique," *Electronics Letters*, Vol.29, No.3, pp.249-250, Feb. 1993.

[28] S. Pookiyaudom, R. Sitdhikorn, "Current-Differencing Band-Pass Filter Realization with Application to High-Ferquency Electronically Tunable Low-Supply-Voltage Current-Mirror-Only Oscillator," *IEEE Trans. on Circuits and Systems- part II*, Vol. 43, No. 12, pp. 832-835, Dec. 1996.

[29] R. Zele, D. Allstot, "Low-Power CMOS Continuous-Time Filters," *IEEE J. of Solid-State Circuits*, Vol. 31, No.2, pp. 157-168, Feb. 1996.

[30] S. Smith, E. Sanchez-Sinencio, "Low Voltage Integrators for High-Frequency CMOS Filters Using Current Mode Techniques," *IEEE Trans. on Circuits and Systems - part II*, Vol. 43, No. 1, pp. 39-48, Jan. 1996.

[31] E. El-Masry, J. Gates, "A novel Continuous-Time Current-Mode Differentiator and Its Applications," *IEEE Trans. on Circuits and Systems - part II*, Vol. 43, No. 1, pp. 56-59, Jan. 1996.

[32] Joung-Chul Ahn, A. Fujii, "Current-Mode Continuous-Time Filters Using Complementary Current Mirror Pairs," *Int. J. Analog Integrated Circuits and Signal Processing*, No. 11, pp. 109-118, 1996.

[33] J. Ramirez-Angulo, E. Sanchez-Sinencio, M. Howe, "Large f_oQ Second Order Filters Using Multiple Output OTAS," *IEEE Trans. on Circuits and Systems - part II*, Vol. 41, No. 9, pp. 587-592, Sept. 1994.

[34] G. Moschytz, A. Carlosena, "A Classification of Current-Mode Single-Amplifier Biquads Based on a Voltage-to-Current Transformation," *IEEE Trans. on Circuits and Systems - part II*, Vol. 41, No.2, pp. 151-156, Feb. 1994.

[35] S. Lee, R. Zele, D. Allstot, G. Liang, "CMOS Continuous-Time Current-Mode Filters for High-Frequency Applications," *IEEE J. of Solid-State Circuits*, Vol. 28, No. 3, pp. 323-328, Mar. 1993.

[36] A. Wyszynski, R. Schaumann, "A Current-Mode Biquadratic Amplitude Equalizer," *Int. J. Analog Integrated Circuits and Signal Processing*, No. 4, pp. 161-166, 1993.

[37] W. Serdijn, "A Low-Voltage Low-Power Current-Mode High-Pass Leapfrog Filter," *Int. J. Analog Integrated Circuits and Signal Processing*, No. 3, pp. 105-112, 1993.

[38] J. Ramirez-Angulo, M. Robinson, E. Sanchez-Sinencio, "Current-Mode Continuous-Time Filters: Two Design Approaches" *IEEE Trans. on Circuits and Systems - part II*, Vol. 39, No. 6, pp. 337-341, June 1994.

[39] R. Brennan, T. Viswanathan, J. Hanson, "The CMOS Negative Impedance Converter," *IEEE J. of Solid-State Circuits*, Vol. 23, No. 5, pp. 1272-1275, Oct. 1988.

[40] J. Haslett, M. Rao, L. Bruton, "High-Frequency Active Filter Design Using Monolithic Nullors," *IEEE J. of Solid-State Circuits*, Vol. SC-15, No.6, pp.955-962, Dec. 1980.

[41] W. Serdijn, M. Broest, J. Mulder, A. van der Woerd, , A. Roermund, "A Low-Voltage Ultra-Low-Power Translinear Integrator for Audio Filter Applications," *IEEE J. of Solid-State Circuits*, Vol. 32, No. 4, pp. 577-581, Apr. 1991.

[42] J. Mulder, A. van der Woerd, W. Serdijn, A. Roermund, "General Current-Mode Analysis Method for Translinear Filters," *IEEE Trans. on Circuits and Systems - part I*, Vol. 44, No. 3, pp. 193-197, Mar. 1997.

[43] Y. Tsividis, "Externally Linear, Time-Invariant Systems and Their Applications to Companding Signal Processor," *IEEE Trans. on Circuits and Systems - part II*, Vol. 44, No. 2, pp. 65-85, Feb. 1997.

[44] D. Perry, G. Roberts, "The design of Log-Domain Filters Based on the Operational Simulation of LC Ladders," *IEEE Trans. on Circuits and Systems - part II*, Vol. 43, No. 11, pp. 763-774, Nov. 1996.

[45] D. Frey, "Log Domain Filtering for RF Applications," *IEEE J. of Solid-State Circuits*, Vol. 31, No. 10, pp. 1468-1475, Oct. 1996.

[46] D. Frey, "Exponential State Space Filters: A Generic Current Mode Design Strategy," *IEEE Trans. on Circ. and Syst - part I*, Vol. 43, No. 1, pp. 34-42, Jan. 1996.

[47] J. Mahattanakul, C. Toumazou, "Modular Log-Domain Filters," *Electronics Letters*, Vol. 33, No. 13, pp. 1130-1131, 1997.

[48] D. Frey, "Log-Domain Filtering: an Approach to Current-Mode Filtering," *IEE Proc. Part G*, Vol.140, No.6, pp. 406-416, Dec. 1993.

[49] D. Frey, "Log Filtering Using Gyrators," *Electronics Letters*, Vol.32, No. 1 pp. 26-28, 1996.

[50] A. Thanachayanont, S. Pookaiyaudom, C. Toumazou, "State-Space Synthesis of Log-Domain Oscillators," *Electronics Letters*, Vol. 31, No. 21, pp. 1797-1799, 1995.

[51] E. Seevinck, "Companding Current-Mode Integrator: A new Circuit Principle for Continuous-Time Monolithic Filters," *Electronics Letters*, Vol.26, No. 24, pp.2046-2047, 1990.

[52] B. Gilbert "Translinear Circuits: An Historical Overview," *Int. J. Analog Integrated Circuits and Signal Processing*, No.9, pp. 95-118, 1996.

[53] R. Wiegerink, H. Ten Pierick, C. Jaspers, W. De Haan, D. De Greef, "Variable-Gamma Circuit for Colour Television Based on the MOS Voltage-Translinear Principle," *Int. J. Analog Integrated Circuits and Signal Processing*, No.9, pp. 189-195, 1996.

[54] R. Wiegerink, "Computer Aided Analysis and Design of MOS Translinear Circuits Operating in Strong Inversion," *Int. J. Analog Integrated Circuits and Signal Processing*, No.9, pp. 181-187, 1996.

[55] A. Andreou, K. Boahen, "Translinear Circuits in Subthreshold MOS," *Int. J. Analog Integrated Circuits and Signal Processing*, No.9, pp. 141-166, 1996.

[56] A. Fabre, "Analysis of Analogue Circuit Implementations from Translinear Blocks," *IEE Proc. Part G*, Vol.138, No.4, pp. 483-491, Aug. 1991.

[57] E. Seevinck, R. Wiegerink, "Generalized Translinear Circuit Principle," *IEEE J. of Solid-State Circuits*, Vol.26, No.8, pp.1098-1102, Aug. 1991.

[58] B. Gilbert, "Translinear Circuits: A proposed Classification," *Electronics Letters*, Vol.11, pp.14-16, 1975.

[59] B. Tellegen, *La Recherche Pour Una Série Complete d'Eléménts de Circuit Ideaux Non-Linéaires*, Rendiconti-Seminario Matematico e Fisico di Milano, Vol.25, pp.134-144, 1954.

[60] A. Payne, C. Toumazou, "Analog Amplifiers: Classification and Generalization," *IEEE Trans. on Circuits and Systems - part I*, Vol. 43, No. 1, pp. 43-50, Jan. 1996.

[61] E. Bruun, "Feedback Analysis of Transimpedance Operational Amplifier Circuits," *IEEE Trans. on Circuits and Systems - part I*, Vol. 40, No. 4, pp. 275-277, Apr. 1993.

[62] J. Millman, A. Grabel, *Microelectronics (second edition)*, McGraw-Hill, 1987.

[63] G. Palumbo, J. Choma Jr., "An Overview of Analog Feedback Part I: Basic Theory," *Int. J. Analog Integrated Circuits and Signal Processing*, Vol. 17, No. 3, 1998.

[64] K. Brehmer, J. Wieser, "Large Swing CMOS Power Amplifier," *IEEE J. of Solid-State Circuits*, Vol. SC-18, No. 6, pp. 624-629, Dec. 1983.

[65] R. Castello, "CMOS Buffer Amplifier," in J. Huijsing, R. van der Plassche, W. Sansen (Ed.) *Analog Circuit Design*, Kluwer Academic Publisher, 1993, pp. 113-138.

[66] G. Caiulo, F. Maloberti, G. Palmisano, S. Portaluri, "Video CMOS Power Buffer with Extended Linearity," *IEEE J. of Solid-State Circuits*, Vol. 28, No. 7, pp. 845-848, July 1993.

[67] F. Eynde, W. Sansen, *Analog Interfaces for Digital Signal Processing Systems*, Kluwer Academic Publisher, 1993.

[68] D. Nelson, S. Evans, *A New Approach to Op Amp Design*, Comlinear Corporation Application Note 300-1, Mar. 1985.

[69] W. Gross, "New High Speed Amplifier Designs, Design Techniques and Layout Problems," in J. Huijsing, R. van der Plassche, W. Sansen, (Eds.) *Analog Circuit Design*, Kluwer Academic Publishers, 1993.

[70] D. Smith, M. Koen, A. Witulski, "Evolution of High-Speed Operational Amplifier Architectures," *IEEE J. of Solid-State Circuits*, Vol. 29, No. 10, pp. 1166-1179, Oct. 1994.

[71] E. Bruun, "Bandwidth Limitations in Current Mode and Voltage Mode Integrated Feedback Amplifiers," *Proc. IEEE ISCAS''95*, Vol. 1, pp. 303-306, Seattle U.S.A., Apr. 1995.

[72] J. Graeme, *Applications of Operational Amplifiers*, McGraw-Hill, 1973.

[73] M. Rao, J. Haslett, "Class AB Bipolar Voltage-Current Convertor," *Electronics Letters*, Vol.14, No.24, pp.762-764, May 1978.

[74] H. J. Huising, *Integrated Circuits for Accurate Linear Analogue Signal Processing*, Delft University Press, 1981.

[75] B. Tellegen, *A General Network Theorem with Applications*, Philips Res. Rep., Vol. 7, pp. 67-72, 1964.

[76] S. Director, R. Rohrer, "The Generalised Adjoint Network and Network Sensitivities," *IEEE Trans. on Circuit Theory*, Vol. CT-16, pp. 318-323, 1969.

[77] L. Magran, A. Arbel, "Current-Mode Feedback Amplifier Employing a Transistorized Feedback Network," *Proc. IEEE ISCAS'94*, London, 1994.

[78] R. Zele, S. Lee, D. Allstot, " A High Gain Current-Mode Operational Amplifier," *Proc. IEEE ISCAS''92*, pp. 2852-2855, San Diego U.S.A., 1992.

[79] A. Arbel, "Comparison Between the Noise Performances of Current-Mode and Voltage-Mode Amplifiers," *Int. J. Analog Integrated Circuits and Signal Processing*, No.7, pp. 221-242, 1995.

Chapter 2

LOW-DRIVE CURRENT AMPLIFIERS

The chapter deals with design aspects of low drive current amplifiers. These amplifiers find applications in on-chip environments, where (load) resistances are strictly controlled by the designer. Therefore, they are not required to deliver output currents higher than the quiescent current of the output branches. As we will show, this class of amplifiers has the strongest potential in terms of low-voltage and high speed capability.

On the other hand, in applications where off-chip loads have to be driven, a current amplifier with high-drive capability is necessary; this item will be dealt with in chapter 3.

The first proposals for current amplifiers were circuit blocks to be used in open-loop configurations. As a common denominator, these solutions were strictly current-mode (i.e., the voltage signal is minimized) and implemented by means of current mirrors [1]-[4]. Therefore, their current gain typically did not exceed 20 dB. These circuits did not satisfy the definition of a current op-amp given in section 1.1. However they found uses in many instruments such as probes, photo-detectors and even in signal processing applications. Examples of this class of current amplifiers are illustrated in Figs. 2.1a and 2.1b [3]-[4].

Both circuits are composed of two (upper and lower) complementary sub-circuits allowing the amplifiers to provide a bipolar output current capability. The translinear loop, made up of transistors M1-M4, suitably sets the input voltage equal to the ground potential under both DC and AC conditions.

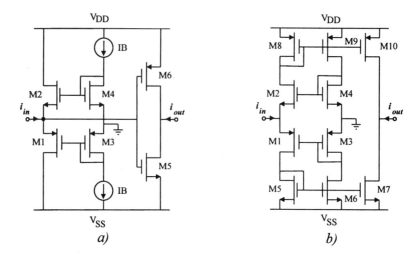

Fig. 2.1. Simple open-loop bidirectional current amplifiers

The input resistance of the circuit in Fig. 2.1a is equal to

$$r_{in} \cong \frac{1}{g_{m1} + g_{m2}} \tag{2.1}$$

and the current gain is

$$A = \frac{i_{out}}{i_{in}} = (g_{m5} + g_{m6})r_{in} = \sqrt{\frac{\beta_{5,6} I_{D5,6}}{\beta_{3,4} I_{D3,4}}} \tag{2.2}$$

where conditions $\beta_5 = \beta_6 = \beta_{5,6}$ and $\beta_3 = \beta_4 = \beta_{3,4}$ for the transconductance gains were assumed.

In the above expressions, as is customary, all transistors are considered to be saturated and body effects are neglected for simplicity.

The circuit in Fig. 2.1b provides a very low input resistance thanks to the additional positive feedback loop in the input stage. The input resistance is given by

$$r_{in} \cong \left[\frac{1}{g_{m1}}\left(\frac{1}{g_{m5}r_{d1}//r_{d5}} + \frac{1}{g_{m3}r_{d3}//r_{d6}}\right)\right] // \left[\frac{1}{g_{m2}}\left(\frac{1}{g_{m8}r_{d2}//r_{d8}} + \frac{1}{g_{m4}r_{d4}//r_{d9}}\right)\right]$$

$$\tag{2.3}$$

The current gain is now set by the aspect ratios of M5 (M8) and M7 (M10)

$$A = \frac{i_{out}}{i_{in}} = (W/L)_7 / (W/L)_5 = (W/L)_{10} / (W/L)_8 \qquad (2.4)$$

Although circuits in Fig. 1.2a and 1.2b provide accurate input resistance and gain, respectively, the quiescent current is controlled well only in the input stage for the former while no current control exists for the latter. This means that most of the performance parameters, such as frequency response, power dissipation, etc., are not well-defined.

Performance with a 3-μm CMOS technology includes a gain ranging typically from 1.1 to 10 and a *GBW* greater than 40 MHz for both circuits.

Actually, a better solution was previously presented by Smith and Sedra in 1987 [5] and is shown in Fig. 2.2.

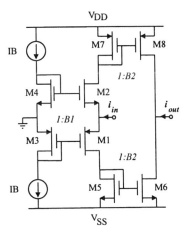

Fig. 2.2. *Mirrored current amplifier*

In this case, the quiescent current in all the branches is set by aspect ratios. More specifically, the DC drain current of transistors M1-M2 and M5, M7 is *B1* times IB, while that of M6 and M8 is *B1·B2* times IB. Typical values of the mirror ratios are 0.1 to 1 for *B1*, and 5 to 10 for *B2*. The input resistance is the same as in (2.1), and the current gain is

$$A = \frac{i_{out}}{i_{in}} = (W/L)_6 / (W/L)_5 = (W/L)_8 / (W/L)_7 = B2 \qquad (2.5)$$

All these amplifiers cannot be identified as current op-amps in the terms stated in section 1.6, because they have only one output terminal. Note that the circuit in Fig. 2.2 was conceived as a current-to-voltage converter to be used in closed-loop applications (by connecting a feedback resistor between the input and the output). However, the poor achievable loop gain prevents the use of this solution in high-performance applications.

In the remaining part of the present chapter, we shall present in detail circuit solutions for the implementation of high-performance feedback stabilized current op-amps (the COA and the VFCOA). We will start our discussion by analyzing different input and output stage implementations by referring to Figs. 1.21 and 1.23, where typical architectures of current amplifiers are shown. These may also include intermediate stages which are voltage amplifiers and, thus, do not require further comments. Then, some design examples of complete amplifiers will be given and critically discussed. Finally, a useful building block related to this class of amplifiers, the *Current Comparator*, will be analyzed and high-performance design solutions provided.

2.1 INPUT STAGES

Current amplifiers make often use of unconventional basic circuits especially to implement the input stage. This is the case for the solutions given in Fig. 2.1a and 2.1b. Unfortunately, some of these unconventional circuits do not provide straightforward operating point control so that, most important performance parameters such as input resistance, transresistance gain, bandwidth, noise, systematic offset, etc., of early current amplifiers presented in literature are not accurately set. These parameters usually require a well-controlled quiescent current in the input stage. Moreover, to avoid systematic offset current and allow input signal generators to be properly biased a well-defined input bias voltage is also required. It is well to remember, here, that current sources are usually made up of transistors in the saturation region.

As already mentioned and illustrated in Figs 1.21 and 1.23, the input stage for both COAs and VFCOAs is the same and can be regarded as a Second Generation Current Conveyor (CCII). In the following section we will briefly introduce this popular building block.

2.1.1 The CCII

The second generation current conveyor (CCII) is, without doubt, the most popular current-mode building block. Since its first presentation in

1970 [6]-[7], many authors have proved its versatility and flexibility in analog circuit design for both linear [8]-[20] and non-linear [21]-[23] applications.

Basically, an ideal CCII is a three terminal device which is labeled X, Y and Z and has the following port relations

$$\begin{bmatrix} v_x \\ i_y \\ i_z \end{bmatrix} = \begin{bmatrix} 0 & 1 & 0 \\ 0 & 0 & 0 \\ \pm 1 & 0 & 0 \end{bmatrix} \cdot \begin{bmatrix} i_x \\ v_y \\ v_z \end{bmatrix} \qquad (2.6)$$

where the + and - signs in the matrix are used for positive (CCII+) and negative type (CCII-) conveyors, respectively. The resistance at terminal Y is ideally infinite, thus no current flows through Y. The current at terminal Z is a replica of current at terminal X. The voltage in X is a replica of the voltage applied to Y. Hence, current in X can be supplied directly through terminal X itself, or can be produced through a copy of the voltage at terminal Y acting across an external impedance connected to X terminal. Therefore, a CCII can be considered as being composed of a voltage follower placed between Y and X, and a positive (CCII+) or a negative (CCII-) current follower which replicates in Z the current flowing through X.

A block diagram of a CCII+ highlighting the two buffers and their related finite internal resistances r_Y, r_X and r_Z is shown in Fig. 2.3.

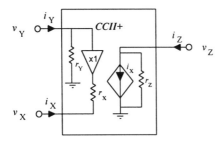

Fig. 2.3. CCII+ block diagram

In addition to current-mode amplifiers, a CCII is well suited for implementing the input stage of a CFOA. To this end, we shall also place an emphasis on solutions for implementing the voltage follower section of a CCII, although this aspect is strictly not so important for current-mode amplifiers. For the current follower section we will commonly assume the simplest topology (i.e., made up of simple current mirrors). Improved current buffers can be implemented using cascode or cascoded topologies. In conclusion, most of the considerations and architectures developed in the following can be directly applied to the design of COA, VFCOA and CFOA input stages.

An important performance parameter of CCIIs used as an input stage is *noise*. Since this subject is not exhaustively treated in existing books, we shall discuss the noise performance of a CCII in greater detail. In general, noise in a CCII can be completely modeled by three noise generators [24], as shown in Fig. 2.4.

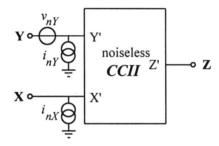

Fig. 2.4. Current Conveyor with equivalent noise generators

Note that noise voltage is only associated with the Y terminal since any noise voltage in series with the X terminal can be directly transferred to the Y terminal. Moreover, any noise current in parallel with the Z terminal can be directly transferred to the X terminal. Hence, terminal X is characterized by only one current noise generator. In a conveyor employed as the input stage of a current-mode amplifier, the Y terminal is always grounded. In this case the noise model can be simplified by omitting the current noise generator i_{nY}. Hence, the noise model in Fig. 2.5, for current amplifier input stages, results

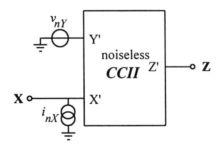

Fig. 2.5. Noise model for a CCII used as current amplifier input stage

For an ideal conveyor whose terminals Y and Z are grounded and X open, the voltage and current noise output powers at terminal X and Z, N_{vX} and N_{iZ}, are, respectively

$$N_{vX} = \overline{v_{nY}^2} \qquad (2.7)$$

$$N_{iZ} = \overline{i_{nX}^2} \qquad (2.8)$$

Class A and class AB CCIIs, suitable for implementing amplifier input stages, will be discussed in sections 2.1.2 and 2.1.3. In particular, a brief analysis of the simplest implementations will first be provided and recent solutions based on high-accuracy configurations will be analyzed and compared in terms of voltage and current gain, A_v and A_i, r_X and noise. Frequency performance will also be evaluated.

2.1.2 Class A Input Stages

A. Basic configurations

The first class A configurations we shall consider are shown in Fig. 2.6, which illustrates very simple solutions for the implementation of a CCII- and a CCII+.

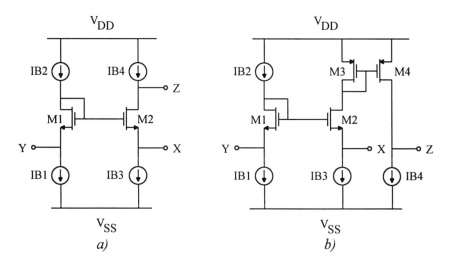

Fig.2.6. *Class A CCIIs a) CCII- b)CCII+*

The voltage follower section is implemented in both circuits with transistor M2 to which an input level shifter (M1) is added. The current follower section is also obtained through M2 which acts in this case as a common gate. In Fig. 2.6b a current mirror is also employed to cause the inversion of the current flow into terminal Z. Although both circuits can operate with a reduced voltage supply (two drain-source saturation voltages and a gate-source voltage are required) they have two main drawbacks. The first is the high voltage-transfer error and the second is the relatively high value of r_X.

Assuming the transistors are saturated and ideally matched, and neglecting body effects, the voltage transfer gain, A_v, A_i, and resistances r_X and r_Z for circuit 2.6a become

$$A_v = \frac{v_X}{v_Y} = \frac{1}{1 + \dfrac{1}{g_{m2} r_{B3}}} \tag{2.9}$$

$$A_i = \frac{i_Z}{i_X} = -1 \tag{2.10}$$

$$r_X \cong \left(\frac{1 + g_{d2} R_{LZ} // r_{B4}}{g_{m2}} \right) // r_{B3} \tag{2.11}$$

$$r_Z \cong \left(\frac{g_{m2}}{g_{d2}} R_{LX} \right) // r_{B4} \tag{2.12}$$

where r_{B3} and r_{B4} are the output resistances of current sources IB3 and IB4 and R_{LZ} is the load resistance at terminal Z. Of course, resistances r_X and r_Z are affected by the load resistances at terminals Z and X, respectively.

For circuit 2.5b we find the same expression as in (2.9) for the voltage transfer gain. The other parameters are

$$A_i = \frac{g_{m4}}{g_{m3}} \tag{2.13}$$

$$r_X \cong \frac{1}{g_{m2}} \tag{2.14}$$

$$r_Z \cong \frac{1}{g_{d4}} // r_{B4} \qquad (2.15)$$

The offset voltage between X and Y is principally due to the mismatch between transistors M1 and M2 and between currents I_{D1} and I_{D2}. It can easily be shown that for both circuits it is

$$V_{os} = V_X - V_Y \cong (V_{T2} - V_{T1}) + \left(\sqrt{\frac{I_{D2}}{\beta_2}} - \sqrt{\frac{I_{D1}}{\beta_1}}\right) \qquad (2.16)$$

The first term in the above equation is related to process tolerances and can only be reduced by careful layout design. The second term depends on design parameters and its main contribution derives from the mismatch between the bias currents I_{D1} and I_{D2}.

The frequency performance of the current-following section relies on the simplicity of the topology. Of course, the current follower in 2.6b has a worst frequency response compared with circuit 2.6a, due to the current mirror M3-M4.

It would be useful to now give an example of noise calculation. The full set of noise generators described in Fig. 2.4 will be evaluated for circuit 2.6b (from these, noise generators for circuit 2.5a can easily be derived). To calculate the noise we use gate-referred (voltage) noise, v_{ni}, for the i-th transistor, and noise current, i_{nBk}, (in parallel) with the k-th bias current generator (see. Fig. 2.7).

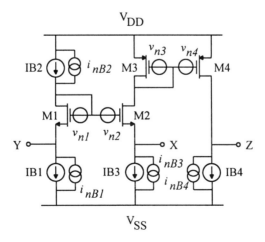

Fig. 2.7. Circuit in Fig. 2.6b with noise generators

Generally, an equivalent input noise generator is obtained by evaluating the output noise and dividing it by the gain of the circuit. For high-gain amplifiers, this leads to the well-known characteristic of input transistors as main contributors to noise. In contrast, CCIIs are made up of two followers, so we expect all the transistors to contribute to noise to a similar extent. The evaluation of the output noise is further complicated by the fact that a CCII is a three-terminal device where terminal X acts as an input and output terminal at the same time. Therefore, care must be taken when evaluating the equivalent output voltage noise at terminal X, to separate the component due to noise voltage from that due to noise current.

It is easy to see that the noise current generator i_{nY} can be found by evaluating the short-circuit current at Y terminal. It is given by

$$\overline{i_{nY}^2} = \overline{i_{nB1}^2} + \overline{i_{nB2}^2} \qquad (2.17)$$

Moreover, by evaluating the open-circuit voltage at X terminal with Y grounded, and the short-circuit current at Z (with terminal X left open) we can compute the remaining generators v_{nY} and i_{nX}. The calculation is simplified assuming the transfer gain of both voltage and current buffers is equal to unity. Note that when we evaluate v_{nY} the component due to i_{nB3} must not be taken into account. According to this observation it results

$$\overline{v_{nY}^2} = \overline{v_{n1}^2} + \overline{v_{n2}^2} + \frac{1}{g_{m1}^2}\overline{i_{nB2}^2} \qquad (2.18)$$

$$\overline{i_{nX}^2} = g_{m4}^2\left(\overline{v_{n3}^2} + \overline{v_{n4}^2}\right) + \overline{i_{nB3}^2} + \overline{i_{nB4}^2} \qquad (2.19)$$

It should be noted that the noise associated with current generator IB2 appears in both (2.17) and (2.18). Hence, a correlation exists between i_{nY} and v_{nY} which has to be taken into account when calculating noise in a system using the current conveyor. Fortunately, the only two equivalent generators really needed for the noise characterization of current amplifier input stages are just those given in (2.18) and (2.19), as stated in section 2.1.1.

The expression of the gate-referred transistor noise ($\overline{v_{ni}^2} = \frac{2}{3}4kT\frac{1}{g_{mi}}df$, where k is Boltzmann's constant and T is the absolute temperature) is minimized by setting large values of g_m. For the circuit in Fig. 2.7 this means setting high the transistor aspect ratios and/or bias currents of M1 and M2.

On the other hand, the noise current is caused by the current mirror M3-M4 and by generators IB3 and IB4. Hence, minimizing this noise means setting low g_m for the current mirror and utilizing low-noise current sources. Unfortunately, this can be achieved using low aspect ratios, since the quiescent currents cannot be set low according to (2.18). Therefore, in practical cases, a trade-off between noise performance and the minimal acceptable supply voltage must be met.

B. Configurations based on the differential stage

Very simple solutions for the implementation of class A CCIIs are also based on the popular differential amplifier and are shown in Figs. 2.8a and 2.8b. Real bias current generators have been drawn to better account for their noise contribution.

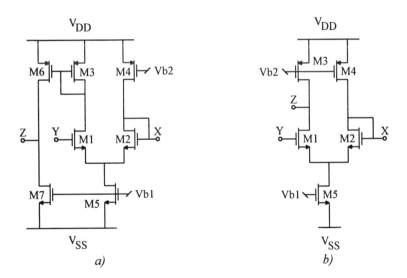

Fig. 2.8. Source-coupled pair CCIIs a) CCII- b) CCII+

Similarly to the previous solutions, both circuits are able to operate under reduced voltage supplies and have the same drawbacks. The voltage transfer gain, A_v, and the resistance, r_X, are given by

$$A_v = \cfrac{1}{1 + \cfrac{1}{\cfrac{1}{2}\cfrac{g_{m1,2}}{g_{d1,2} + g_{d3}}}} \qquad (2.20)$$

$$r_X \cong \frac{2}{g_{m1,2}} \qquad (2.21)$$

It is clear that the current transfer gain between terminal X and Z is the same as the gain expressed in (2.10) and (2.13). In addition resistance r_Z is about equal to $1/g_{d3}$ and $1/(g_{d6}+g_{d7})$ for the circuits in Fig. 2.8a and 2.8b, respectively.

The offset voltage between X and Y, is principally caused both by mismatch between transistors M1 and M2 and that existing between currents I_{D4} and I_{D5} (ideally we should have $I_{D5} = 2\ I_{D4}$). It can easily be shown that the relationship (2.16) for the offset still holds.

Finally, the equivalent noise generators of the CCII in Fig. 2.8a are evaluated. Since terminal Y is the gate of a MOS transistor, noise current i_{nY} is equal to zero. Noise voltage v_{nY} accounts for the contribution of transistors M1 and M2. It is given by

$$\overline{v_{nY}^2} \cong 2\overline{v_{n1,2}^2} \qquad (2.22)$$

Noise current i_{nX} is given by

$$\overline{i_{nX}^2} \cong g_{m6}^2(\overline{v_{n3}^2} + \overline{v_{n6}^2}) + g_{m7}^2(\overline{v_{n5}^2} + \overline{v_{n7}^2}) + g_{m4}^2\left(\frac{g_{m6}}{g_{m3}}\right)^2\overline{v_{n4}^2} + g_{m5}^2\left(\frac{g_{m6}}{g_{m3}}\right)^2\overline{v_{n5}^2} =$$

$$= 3g_{m3,4,6}^2\overline{v_{n3,4,6}^2} + g_{m5}^2\overline{v_{n5}^2} + g_{m7}^2\left(\overline{v_{n5}^2} + \overline{v_{n7}^2}\right) \qquad (2.23)$$

where transistor M5 contributes to the equivalent noise with two terms, since there are two different paths from the gate of M5 to the output. The first is through M7 and the second through M1, M3 and M6. The noise generators of the circuit in Fig. 2.8b can be obtained with some simplifications from those of the circuit shown in Fig. 2.8a.

C. Improved configurations

In this section we shall present some improved CCII solutions. Only positive conveyors will be treated since the main parameters are about equal for both CCII+ and CCII-. The first improvement on the basic structure involves the reduction of r_X, which improves both the voltage and current transfer gains. In order to do this, a source follower stage, M9, is added in a closed loop configuration to the source-coupled pair as shown in Fig. 2.9a, [25]. In this way, according to the expressions reported in Table 2.1, the resistance at terminal X is reduced by about a stage gain, and the differential stage achieves full transconductance thanks to the current mirror active load M3-M4. In contrast, the circuit shown in Fig. 2.9b [26], provides a simpler topology with the same A_v and r_X (see Table 2.1) as that in Fig. 2.9a. Moreover, a similar low-voltage capability to circuits in Fig. 2.8 is also achieved.

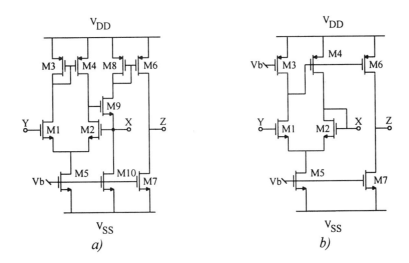

Fig. 2.9 a) and b). Improved CCIIs

Drawbacks of the circuit in Fig. 2.9b when compared to the one in Fig. 2.9a are a larger offset voltage caused by mismatches in the biasing circuits and a reduced linearity in the voltage transfer gain. Indeed, the tolerances and channel length modulation effects in the current mirrors of the biasing circuits (not shown in the figure), mean that current I_{D1} is not kept exactly equal to $\frac{I_{D5}}{2}$. Consequently, according to (2.16), a large offset may occur.

Moreover, the drain-source voltage of M5 and hence current I_{D5} depend on the input signal, thus affecting circuit performance.

Table 2.1.
Main electrical parameters of the CCIIs in Fig. 2.9

Circuit ref.	A_v	r_X
Fig. 2.9a	$\dfrac{1}{1+\dfrac{1}{\dfrac{g_{d1,2}+g_{d4}}{g_{m1,2}}}}$	$\dfrac{1}{g_{m9}}\dfrac{g_{m1,2}}{g_{d1,2}+g_{d4}}$
Fig. 2.9b	$\dfrac{1}{1+\dfrac{1}{\dfrac{g_{d1,2}}{g_{m1,2}}}}$	$\dfrac{g_{d1,2}+2g_{d4}}{g_{m1,2}g_{m4}}$

To reduce offset and non-linearity for the Current Conveyor in Fig. 2.9b, the biasing, MC1-MC3, in Fig. 2.9c was used.

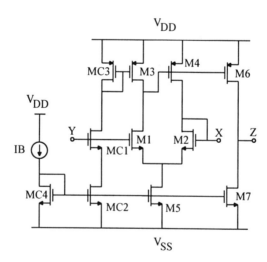

Fig. 2.9c. *CCII in Fig. 2.9b with improved biasing*

If MC1 and MC3 are matched to M1 and M3, respectively, and the aspect ratio of MC2 is half that of M5, the common drain transistor MC1 can keep the drain-source voltage of MC2 equal to that of M5, regardless of signal V_Y. Therefore, the current in MC3 is forced to be dynamically matched to that in M5. Moreover, since transistors M3 and M4 have the same currents and aspect ratios, the drain-source voltage of M3 is equal to that of MC3. As a result, current mirror MC3-M3 is not affected by channel-length modulation. Simulations showed that THD can be improved by 10 dB with this technique [28].

The equivalent input noise at node Y of the circuit in Fig. 2.9a and 2.9b is slightly different from that of the circuit in Fig. 2.8. With the circuit in Fig. 2.9a M3-M4 contribute to noise, whilst the noise due to M9 is suppressed by the local feedback formed by M2 and M9. The equivalent voltage noise is

$$\overline{v_{nY}^2} \cong 2\overline{v_{n1,2}^2} + 2\left(\frac{g_{m3,4}}{g_{m1,2}}\right)^2 \overline{v_{n3,4}^2} \qquad (2.24)$$

On inspection, it is easy to see that only transistors M6, M7, M8 and M10 contribute to the noise current, while the noise of M5 appears as a common-mode signal.

The current noise is

$$\overline{i_{nX}^2} \cong g_{m6}^2 \overline{v_{n6}^2} + g_{m7}^2 \overline{v_{n7}^2} + g_{m6}^2 \overline{v_{n8}^2} + g_{m10}^2 \overline{v_{n10}^2} + g_{m10}^2 \overline{v_{n5}^2} + g_{m7}^2 \overline{v_{n5}^2} = \qquad (2.25)$$

$$= 2g_{m6,8}^2 \overline{v_{n6,8}^2} + g_{m7}^2 \left(\overline{v_{n7}^2} + \overline{v_{n5}^2}\right) g_{m10}^2 \left(\overline{v_{n10}^2} + \overline{v_{n5}^2}\right)$$

Similarly, for the circuit in Fig. 2.9b we have

$$\overline{v_{nY}^2} \cong 2\overline{v_{n1,2}^2} + \left(2\frac{g_{m3}}{g_{m1,2}}\right)^2 \overline{v_{n3}^2} + \left(\frac{g_{m5}}{g_{m1,2}}\right)^2 \overline{v_{n5}^2} \qquad (2.26)$$

$$\overline{i_{nX}^2} \cong g_{m3}^2 \overline{v_{n3}^2} + g_{m5}^2 \overline{v_{n5}^2} + g_{m6}^2 \left(\overline{v_{n4}^2} + \overline{v_{n6}^2}\right) + g_{m7}^2 \left(\overline{v_{n7}^2} + \overline{v_{n5}^2}\right) \qquad (2.27)$$

Further improvements can be achieved by increasing the open-loop gain of the voltage buffer. Such an approach, which involves two gain stages, was recently proposed in [27]. However, it does not completely fulfill its aims. Indeed, the solution requires quite a complex circuit implementation and uses various current mirrors which cause inaccuracy, especially for low voltage operations.

A simple and more accurate two gain stage topology using simple current mirrors is illustrated in Fig. 2.10 [28]. In the schematic in Fig. 2.10a, block VA is a voltage amplifier which increases the loop gain, and block CB is a current buffer which provides high impedance driving for the current mirror M4-M6. A feasible implementation of the circuit in Fig. 2.10a is shown in Fig. 2.10b where block VA is the common source amplifier M8-M9 and transistor M10 has a double function. It implements block CB and preserves the high gain of the common source stage.

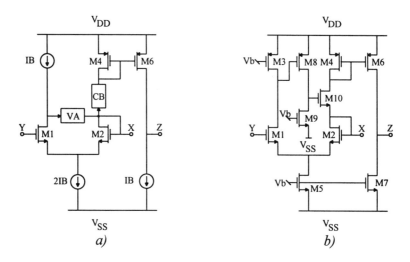

Fig. 2.10. Two stage enhanced CCII-:
a) simplified schematic, b) detailed schematic

Following common simplifications, the transfer gain from Y to X and the resistance at terminal X are given by

$$A_v \cong \cfrac{1}{1 + \cfrac{1}{\cfrac{1}{2} \cfrac{g_{m1,2} g_{m8}}{g_{d3} g_{o2}}}} \qquad (2.28)$$

$$r_x \cong \frac{1}{\left(\dfrac{g_{m1,2}}{2} + g_{m10}\right) \dfrac{1}{2} \dfrac{g_{m1,2} g_{m8}}{g_{d3} g_{o2}}} \qquad (2.29)$$

where $g_{o2} = g_{d8} + g_{d9}$ is the equivalent conductance at the output of the common source.

Finally, it can be demonstrated that noise equations (2.26) and (2.27) also hold for this circuit. Hence, this solution provides more accurate performance with the only expense of a reduced low-voltage capability.

D. Simulated results

To compare the performance of most of the previously discussed circuits (in Figs. 2.8, 2.9 and 2.10), some simulated results are given. SPICE models using a standard 1.2-μm n-well CMOS process were used. Apart from transistor M5 whose aspect ratio was set equal to 20/2, the aspect ratios for *n*-channel and *p*-channel transistors were all set to 10/2 and 30/2, respectively. The power supply was set to 5 V and the bias current $I_{D5} = 2 I_{D3,4}$ was set to 20 μA. A load capacitor, C_L, with 2 pF at terminal X was considered. Simulated and calculated results regarding transfer gain and resistance at terminal X are summarized in Table 2.2. The values expected from the previous equations are in good agreement with the simulated ones. It is obvious that r_X is reduced by about two orders of magnitude starting from the circuits in Figs. 2.8a to 2.9a (or 2.9b) and then to 2.10.

Noise performance was also validated by simulations. Results are shown in Table 2.3, showing the equivalent input noise voltage and current evaluated at a much higher frequency than the flicker noise corner.

Although the circuits were not designed to obtain the maximum bandwidth, we also simulated the frequency response of the voltage transfer gain to compare their frequency performance. The Bode diagrams are shown in Fig. 2.11. Except for the circuit in Fig. 2.8a, frequency compensation was provided to guarantee frequency stability. The low performance of the circuit in Fig. 2.8a is apparent as are the accurate transfer gains of the circuits in Figs. 2.9b and 2.10. The frequency response of the current transfer gain between terminal X and Z with $R_Z = 100$ Ω is shown in Fig. 2.12.

Table 2.2.
Values of A_v and r_X for class A CCIIs

Circuit ref.	A_v (simulated)	A_v (calculated)	r_X (simulated)	r_X (calculated)
Fig. 2.8a	0.974	0.972	32 kΩ	28 kΩ
Fig. 2.9a	0.986	0.986	207 Ω	195 Ω
Fig. 2.9b	0.994	0.994	310 Ω	278 Ω
Fig. 2.10	0.999	1.000	3 Ω	2 Ω

Table 2.3.
Noise for class A CCIIs

Circuit ref.	$\sqrt{\overline{v_{nY}^2}}\ [nV/\sqrt{Hz}]$	$\sqrt{\overline{i_{nX}^2}}\ [pA/\sqrt{Hz}]$
Fig. 2.8a	17	2.3
Fig. 2.9a	25	1.9
Fig. 2.9b	34	2.3
Fig. 2.10	37	2.4

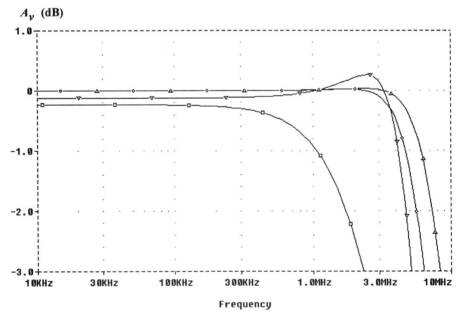

Fig. 2.11. *Voltage transfer gain for the circuits in Figs 2.8 to 2.10*

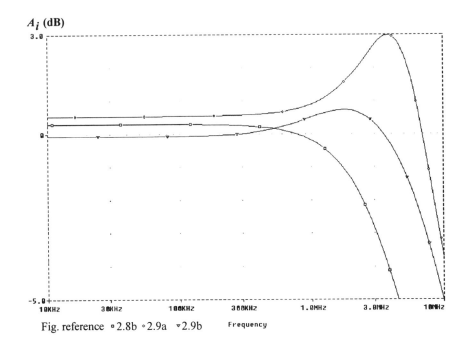

Fig. reference ▫2.8b ∘2.9a ▾2.9b

Fig. 2.12. *Current gain for the circuits in Figs 2.8 to 2.9*

2.1.3 Class AB Input Stages

Although class A input stages provide high accuracy, good frequency response, and have the greatest potential for low voltage operations, class AB topologies are often preferred for their superior Slew Rate performance. Indeed, the output swing is no longer limited by the quiescent current. The latter property contributes positively to the better signal to noise ratios provided by class AB topologies [24].

A. Configurations

Two of the most commonly used circuit implementations for the voltage follower in a class AB input stage are shown in Fig. 2.13.

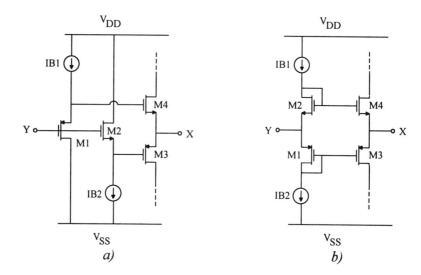

Fig. 2.13. Two class AB voltage follower implementations *a)* and *b)*

The two circuits provide well-controlled bias currents and output voltages, have the same output impedance, and achieve the same drive capability. By setting IB1 = IB2 = IB and the transistor aspect ratios as follows

$$\frac{\left(\frac{W}{L}\right)_3}{\left(\frac{W}{L}\right)_1} = \frac{\left(\frac{W}{L}\right)_4}{\left(\frac{W}{L}\right)_2} = n \qquad (2.30)$$

transistors M1, M3, and M2, M4 acquire the same gate-source voltages, with the current in M3, M4 being set to $n \cdot$IB. Assuming transistors with the same bulk-source voltages, M1-M4 provide a translinear loop which accurately reflects the DC voltage of terminal Y on terminal X. In fact, process tolerances do not appreciably modify the operating point.

The input resistance at terminal X for both circuits is given by (2.1). To achieve a very low input resistance in most applications, rather large transistor aspect ratios and/or high bias currents have to be used for transistors M3-M4. The circuit in Fig. 2.13a exhibits an infinite Y resistance, while the Y resistance in Fig. 2.13b is the parallel combination of the output resistances of current sources IB1 and IB2.

Note that to provide accuracy, low offset voltage and good linearity, the circuit in Fig. 2.13a requires a twin tub process. Otherwise, *n*-channel and *p*-

channel transistors will have different bulk-source voltages. In contrast, the circuit in Fig. 2.13b is not affected by this problem since either transistors M1-M3 or M2-M4 must be placed in a well. A minor drawback is that current IB1 and IB2 in Fig. 2.13b have to be accurately matched in order to avoid systematic input offset current.

To conclude, both circuits provide class AB operation, have a simple implementation and quite similar performance. However, the circuit in Fig. 2.13b is used more frequently because it needs a standard single-well CMOS process.

To improve performance, feedback can be employed, as shown in Fig. 2.14. Here a differential amplifier is combined with the circuit in Fig. 2.13b to increase the resistance at terminal Y, decrease the resistance at terminal X, and reduce offset voltage and gain error.

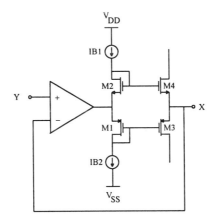

Fig. 2.14. Class AB voltage follower using feedback

The complete schematic of class AB CCII+ and CCII- implemented with the voltage follower given in Fig. 2.13b are shown in Figs 2.15. Two cascoded complementary current mirrors are adopted in the CCII+ in Fig. 2.15a, while two complementary folded-cascode structures is the natural choice for the CCII-. For the latter, current generator IB4 (IB3) provides the quiescent current in M4 (M3) and sets it in M6 (M5).

It is worth noting that despite the class AB voltage follower, only the first circuit (CCII+) provides a true class AB configuration for the current follower as well. Only this circuit is, in fact, able to deliver an output current which is not limited by any constant value. As a result, to achieve high slew rate values the CCII+ in Fig. 2.15a must be used.

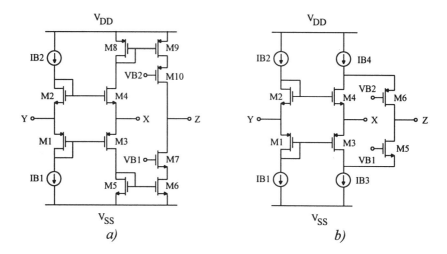

Fig. 2.15. *Class AB CCIIs: a) CCII+ and b) CCII-*

Under the usual assumptions, the voltage transfer gain, A_v, for both circuits is

$$A_v = \frac{v_X}{v_Y} = \frac{1}{1 + \dfrac{g_{d3} + g_{d4}}{g_{m3} + g_{m4}}} \qquad (2.31)$$

while the expression of resistance r_X has already been reported in (2.1).

Of course, the current transfer gain depends on the mirror ratios of transistors M5-M6 and M8-M9 for the CCII+ while it is accurately set to 1 in the CCII-. The expression of resistance r_Z is typical of cascode structures.

Noise v_{nY}, in both circuits is given by

$$\overline{v_{nY}^2} = \left(\frac{g_{m3}}{g_{m3} + g_{m4}}\right)^2 \left(\overline{v_{n2}^2} + \overline{v_{n4}^2} + \frac{1}{g_{m2}^2}\overline{i_{nB2}^2}\right) +$$

$$+ \left(\frac{g_{m4}}{g_{m3} + g_{m4}}\right)^2 \left(\overline{v_{n1}^2} + \overline{v_{n3}^2} + \frac{1}{g_{m1}^2}\overline{i_{nB1}^2}\right) \qquad (2.32)$$

where i_{nB1} and i_{nB2} are the equivalent noise currents of IB1 and IB2, respectively.

Current generators, i_{nX}, for the CCII+ and CCII- in Fig. 2.19, are given, respectively

$$\overline{i_{nX}^2} = g_{m6}^2\left(\overline{v_{n5}^2} + \overline{v_{n6}^2}\right) + g_{m9}^2\left(\overline{v_{n8}^2} + \overline{v_{n9}^2}\right) \tag{2.33a}$$

$$\overline{i_{nX}^2} = \overline{i_{nB3}^2} + \overline{i_{nB4}^2} \tag{2.33b}$$

B. Simulations

The two circuits in Fig. 2.15 were simulated using the model parameters of a 1.2-μm CMOS process. The aspect ratio of the n-channel transistors is 10/2 while that of the p-channel transistors is 30/2 and the bias current in all the branches was set to 10 μA. Moreover, the two bias voltages, VB1 and VB2, were set to 2.5 V (i.e., equal to $V_{DD}/2$). Finally, a 2-pF load capacitance at terminal X was assumed.

Resistances at terminal X and terminal Z were found to be 5.8 kΩ and 16 MΩ for the circuits in Fig. 2.15a and 5.9 kΩ and 19.5 MΩ those in Fig. 2.15b. Their noise performance is summarized in Table 2.4. Real bias current sources were used with the same aspect ratios reported above. Of course, noise voltage for both circuits has the same value. The second solution is noisier than the first one, despite the simpler equation (2.32b). This is because the noise contribution of a current source (IB3 and IB4) is usually larger than that from a couple of transistors, e.g. a current mirror. Moreover, current IB3 (IB4) was set twice as high as IB1 (IB2), leading to wider transistors which further increase the noise current.

Table 2.4.
Noise in class AB CCIIs

Circuit ref.	$\sqrt{\overline{v_{nY}^2}}$ $\left(nV/\sqrt{Hz}\right)$	$\sqrt{\overline{i_{nX}^2}}$ $\left(pA/\sqrt{Hz}\right)$
Fig. 2.15a	15	1.9
Fig. 2.15b	15	1.5

To allow comparison between class A and class AB topologies, the same bias currents used in Fig. 2.6 were set for the circuits in Fig. 2.15. Comparing Table 2.3 with Table 2.4, the advantage of class AB circuits over their class A counterparts is not apparent. However, in a true class AB topology (like the one in Fig. 2.15a) the maximum output current is much higher than the quiescent current. This means that assuming the same maximum output current for class A and class AB stages, the latter can be designed using smaller aspect ratios and bias currents thus reducing the noise current although at the expense of an increase in noise voltage. Fortunately, increasing noise voltage is of minor importance in a current amplifier, since the effect of noise voltage can be minimized by setting the feedback resistance R_1 high, as according to 1.37.

The frequency response of the voltage transfer gain for both circuits is shown in Fig. 2.16. It gives a low-frequency transfer error of about -0.2 dB and a -3 dB frequency of 15 MHz (due to the output time constant $r_X C_L$).

The current transfer gain between terminal X and Z with R_Z = 100 Ω for both circuits is shown in Fig. 2.17. As expected, a better accuracy is observed in the second circuit.

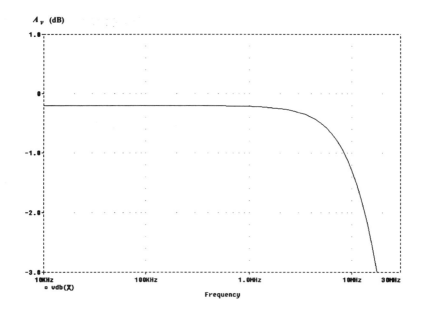

Fig. 2.16. Voltage transfer gain for the circuits in Fig 2.15

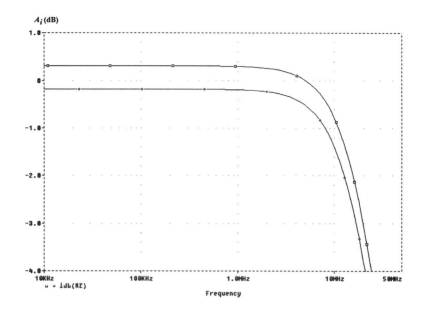

Fig. 2.17. *Current transfer gain for the circuits in Fig 2.15:*
□ *CCII+ ;* ◊ *CCII-*

2.2 CLASS 'A' CURRENT OUTPUT STAGES

We have already stated that the suitable current output stage (COS) configuration for a COA is a differential-output transconductance stage, whereas that for a VFCOA is a CCII-. It is important to emphasize that the design of a COS is not so straightforward as a voltage-mode output stage. Indeed, while in voltage amplifiers the same output terminal drives the load and provides feedback, a current amplifier has two different outputs, one for the feedback network and one for the load. The implementation of high-performance COSs requires architectures operating wholly within the feedback loop, so as to provide a load current derived directly from the feedback current. In this manner, COSs can match their voltage counterparts in terms of accuracy, speed and output resistance.

As we shall see, attaining such performance requires class A topologies which, in turn, means placing a limitation on amplifier current drive capability. This section mainly addresses the design of class A current output stages, whereas class AB COS design will be one of the issues dealt with in chapter 3.

2.2.1 Output Stages for COAs

A. Configurations

The simplest solution for implementing a transconductance output stage is the well-known source-coupled pair with constant current loads. The circuit can be driven by a differential or single-ended voltage source. For the latter, one of the two input terminals is connected to a constant voltage, as illustrated in Fig. 2.18.

The stage provides an equivalent transconductance given by

$$G_m = \frac{i_o^+}{v_i} = -\frac{i_o^-}{v_i} = \frac{1}{2} g_{m1,2} \qquad (2.34)$$

where ideal matching conditions have been assumed for transistors M1 and M2. However, if ideal current sources are used, a mismatch between M1 and M2 causes each transconductance g_{mi} to vary. However, the small-signal current i_o^+ remains equal to $-i_o^-$. This last condition guarantees accurate closed-loop performance.

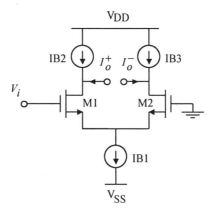

Fig. 2.18. Current output stage implemented with a differential couple

The differential and common-mode transconductances, G_{md} and G_{mc} are

$$G_{md} = \frac{i_o^+ - i_o^-}{v_i} = g_{m1,2} \qquad (2.35a)$$

$$G_{mc} = \frac{i_o^+ + i_o^-}{2v_i} = \frac{1}{2r_{B1}} \qquad (2.35b)$$

where r_{B1} is the output resistance of current source IB1.

Since the common-mode rejection ratio depends only on the output stage, it is given by

$$CMRR = 2g_{m1,2} r_{B1} \qquad (2.36)$$

We see that a very high CMRR can be achieved by using current sources with very high output resistances [29].

The differential and common-mode output resistances, including the output resistance of current sources, are

$$r_{od} = 2r_{d1,2} // r_{B2,B3} \qquad (2.37a)$$

$$r_{oc} = r_{B2,B3} // (2g_{m1,2} r_{d1,2} r_{B1}) \qquad (2.37b)$$

With simple current generators (i.e., implemented with single transistors) resistances r_{B2} and r_{B3} are similar to $r_{d1,2}$. Consequently their values determine both r_{od} and r_{oc}. In section 1.7.1 we showed that the common mode resistance places an upper limit on the maximum closed-loop output resistance achievable with feedback. This means high values of r_{oc} are desirable, which can be obtained from a cascode current source. However, this option also limits the low-voltage capability which is the main advantage behind this simple transconductance stage.

In section 1.7, we showed how the output stage noise contributes to the overall noise performance of a current amplifier. We modeled noise generated in COS with a noise generator at the output of the amplifier. For the circuit in Fig. 2.18, noise is mainly due to the three bias current generators, since noise from M1 and M2 is reduced by the loop gain in a real configuration. Therefore, the output noise current is

$$\overline{i_{no}^2} = \overline{i_{nB1}^2} + \overline{i_{nB2}^2} + \overline{i_{nB3}^2} \qquad (2.38)$$

By interconnecting two complementary source-coupled pairs, we obtain the *floating current source*, first reported in [30] by Arbel and illustrated in Fig. 2.19.

Again, the circuit can be driven by a differential or single-ended voltage source. The equivalent transconductance is given by

$$G_m = \frac{i_o^+}{v_i} = -\frac{i_o^-}{v_i} = \frac{1}{2}\left(g_{m1,2} + g_{m3,4}\right) \qquad (2.39)$$

Assuming IB1 is equal to IB2, to set equal transconductance for both source-coupled pairs, the usual condition for *n*- and *p*-channel transistor aspect ratios has to be chosen. It is

$$\frac{\left(\frac{W}{L}\right)_n}{\left(\frac{W}{L}\right)_p} = \frac{\mu_p}{\mu_n} \qquad (2.40)$$

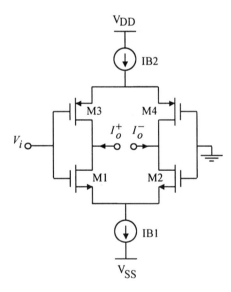

Fig. 2.19. Arbel current output stage (floating current source)

The output range in which the transconductance, G_m, varies by less than 10% is [30]

$$\left|i_{oMAX}^+\right| = \left|i_{oMAX}^-\right| < \frac{I_{B1,2}}{4} \qquad (2.41)$$

The differential and common-mode transconductances, G_{md} and G_{mc} are

$$G_{md} = \frac{i_o^+ - i_o^-}{v_i} = g_{m1,2} + g_{m3,4} \qquad (2.42a)$$

$$G_{mc} = \frac{i_o^+ + i_o^-}{2v_i} = \frac{1}{2r_{B1}} + \frac{1}{2r_{B2}} \qquad (2.42b)$$

where r_{B1} is the output resistance of current source IB1. The common-mode rejection ratio is

$$CMRR = 2 \frac{g_{m1,2} + g_{m3,4}}{\frac{1}{r_{B1}} + \frac{1}{r_{B2}}} \qquad (2.43)$$

Again, a very high *CMRR* can be obtained using current sources with very high output resistances.

The differential and common-mode output resistances, with real bias current sources, are

$$r_{od} = 2(r_{d1,2} // r_{d3,4}) \qquad (2.44a)$$

$$r_{oc} = 2\left[\left(g_{m1,2} r_{d1,2} r_{B1}\right) // \left(g_{m3,4} r_{d3,4} r_{B2}\right)\right] \qquad (2.44b)$$

Finally, the output noise is given by

$$\overline{i_{no}^2} = \overline{i_{nB1}^2} + \overline{i_{nB2}^2} \qquad (2.45)$$

B. Simulations

To compare the performance of the previously discussed circuits (in Figs 2.18 and 2.19), simulated results are given using SPICE and the models of a 1.2-μm *n*-well CMOS process. In both circuits, the aspect ratios for *n*-

channel and *p*-channel transistors were set to 10/1.2 and 30/1.2, respectively and real current generators were used. The power supply was set to 5 V and bias currents IB1 and IB2 to 20 µA. Simulated results for differential and common-mode transconductances and output resistances are shown in Table 2.5.

Table 2.5
Performance of the current output stages

Circ. Ref.	G_{md} (µA/V)	G_{mc} (µA/V)	r_d (kΩ)	r_c (kΩ)
2.18	100	1.2	556	800
2.19	180	1.9	696	20 10³

Using cascode bias current generators the common mode transconductance gain of the circuit in Fig. 2.19 is reduced to 0.04 µA/V.

2.2.2 Output Stages for VFCOAs

The topologies of CCII- discussed in section 2.1 can be used to implement the output stage of a VFCOA. However, to give output stages accurate performance, current mirrors should be avoided. From this point of view, the best solution is shown in Fig. 2.15b where current follower is operated in class A, although the circuit was classified as class AB CCII.

As mentioned before, class A output stages allow the load current to be directly derived from the feedback signal, thus providing linear and accurate closed-loop gain. Additional considerations regarding VFCOAs can be found in section 2.3.2 which discusses practical implementations.

2.3 DESIGN EXAMPLES

In this section we present a number of configurations of low-drive current amplifiers. Specifically, we will discuss three COA and two VFCOA architectures which appropriately arrange the previously described basic building blocks. The examples chosen mainly focus on the capabilities of current-mode amplifiers in terms of low voltage, slew rate and bandwidth. Of course, not all the best features can be provided at the same time meaning a trade-off is necessary.

For typical signal-processing requirements, a two-stage current amplifier configuration is usually adequate. It is made up of the cascade of a transimpedance amplifier with a current output stage, the transimpedance amplifier being implemented with a simple or a cascode gain stage. Since the current output stage is loaded with quite low feedback resistances, dominant-pole compensation is usually adopted to provide stability.

2.3.1 COA Configurations

A. First Design

The first COA presented here was proposed by Bruun [31] and is designed to operate with a low supply voltage and reduced power dissipation. The circuit is very simple being made up of a common gate transistor (which implements the input current follower), a current mirror and a source-coupled pair (implementing the output stage) as illustrated in Fig. 2.20a. With this arrangement, the circuit only requires a gate-source voltage plus two drain-source saturation voltages to operate properly, provided that simple current mirrors are used to implement the bias current generators. Therefore, a supply voltage of 1.2 V is allowed even with standard CMOS processes having threshold voltages of 0.9 V. However, to achieve a high open-loop gain, (low-voltage) cascoded current mirrors must be used, as illustrated in the detailed schematic of Fig. 2.20b. This choice requires a higher bias voltage and in the actual design, a single 1.5-V supply is used.

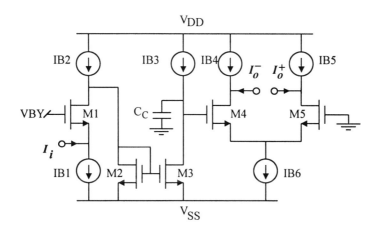

Fig. 2.20a. *Simplified schematic of the first COA solution*

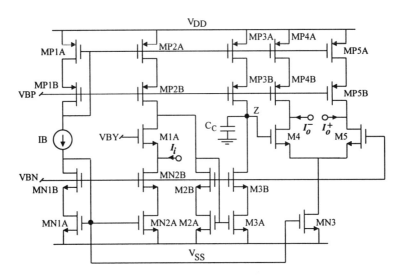

Fig. 2.20b. Detailed schematic of the first COA solution

The circuit was fabricated with a 2-μm technology. All transistors have minimum channel length and the aspect ratio of all *n*-transistors is 10. The aspect ratio of *p*-transistors is 30 except for transistors MP2 (60) and MP4-MP5 (15). The bias current, IB, is 5 μA and C_C is the compensation capacitor which, with the associated resistance r_Z, determines the dominant pole of the circuit.

Of course, a class A CCII+ used as input stage limits the slew rate performance, but it is necessary to achieve low-voltage capability.

The main performance is summarized in Table 2.6 which reports a comparison with other two COA designs discussed in the following.

Table 2.6.
Measured performance of COAs

Circuit ref.	Fig. 2.20	Fig. 2.21	Fig. 2.22
Technology	2 μm	2 μm	1.2 μm
V_{DD}-V_{SS}	1.5 V	3 V	6 V
DC Power	$\cong$ 40 μW	$\cong$ 90 μW	4.5 mW
Gain	94 dB	96 dB	70 dB
GBW	65 MHz	128 MHz	210 MHz
M_ϕ	40 °	---	45 °

B. Second Design

The second solution combines high slew rate and bandwidth [32]. To achieve the first feature a class AB CCII is employed as input stage. Moreover, since this CCII topology limits the supply voltage to about 3.5 V, the differential floating current source in Fig. 2.19 can profitably be employed without sacrificing low-voltage capability while accurate closed-loop performance is ensured. The detailed schematic of the circuit is shown in Fig. 2.21. It was implemented in a 2-µm process. All *n*-channel and *p*-channel transistors have identical aspect ratios equal to 10 and 30, respectively. Main amplifier performance is summarized in the second column of Table 2.6.

In these two designs the authors neglected the input pole. The input pole is due to the open-loop input resistance of the COA with its associated input capacitance. Since the order of magnitude of the input resistance is $1/g_m$ (for both the first and second designs) and the input capacitance amounts to at least the drain-bulk capacitance of a current source transistor, the input pole usually has the same magnitude as the poles due to the current mirrors (which were the only poles considered in the design steps). Consequently, neglecting the input pole can lead to stability problems, especially with relatively high *GBW*, as is evidenced in the poor phase margins given in Table 2.5.

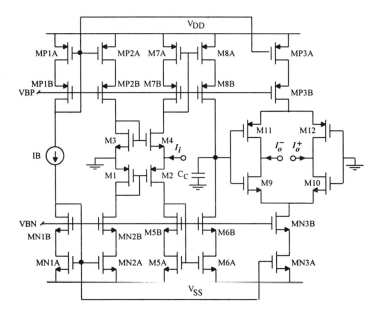

Fig. 2.21. Detailed schematic of the second COA solution

C. Third Design

To achieve even better frequency performance, the configuration shown in Fig. 2.22a was developed [33]. The use of current mirrors is avoided and a common gate transistor (M2) is employed instead. Thanks to the use of an auxiliary differential amplifier, a local feedback in the input stage is provided which lower the input resistance. Of course, to improve the frequency performance the gain-bandwidth product of the input local feedback must be higher than in the main amplifier. Under this condition the main loop gain becomes

$$T(s) = \frac{G_m r_Z}{(1+sr_Z C_C)\left(1+s\dfrac{C_A}{g_{m2}}\right)\left(1+s\dfrac{r_{in}}{1+A}C_{in}\right)} \quad (2.46)$$

where G_m is the transconductance of the output stage, r_Z is the equivalent resistance at node Z, C_A and C_{in} are the equivalent capacitances at node A and the input terminal, respectively. The detailed schematic of the circuit is illustrated in Fig. 2.22b, where the auxiliary amplifier is implemented with the differential stage M5-M8. The simple differential output stage was replaced by a folded-cascode transconductance amplifier. This, however, introduces another non-dominant pole into the loop gain which is of the same order as the one in equation (2.45). Consequently, it has to be accounted for when preserving the phase margin. Main amplifier performance is summarized in the last column of Table 2.6.

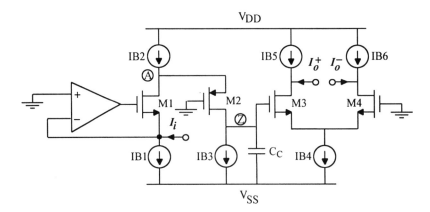

Fig. 2.22a. Simplified schematic of the third COA solution

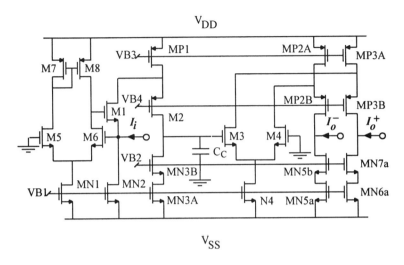

Fig. 2.22b. Detailed schematic of the third COA solution

As already noted, in the first two designs the authors neglected the input pole which can be of the same magnitude as those due to current mirrors. In the third design, the input resistance is greatly reduced by the input feedback loop, but an extra pole must be considered if the *GBW* of the feedback loop is not much higher than that of the main loop, as mentioned before. These considerations mean that the designs discussed should have been compensated with higher compensation capacitors. Therefore, the potential *GBW*s seems to be slightly lower than those reported in Table 2.5.

2.3.2 VFCOA Configurations

A. First design

To the author's knowledge, no integrated version of class A VFCOAs have been reported in literature. Therefore, only a simulated design of a VFCOA example where no parameter optimization has been performed, will be presented.

The circuit is shown in Fig. 2.23 and is composed of an input CCII+ and an output CCII-. The input class AB conveyor provides high slew rate performance, while the output CCII- features a class A current following behavior. This allows accurate closed loop performance to be achieved since the input and the output current are closely and linearly related even in open-loop conditions.

All n-channel and p-channel transistors were set to 10/1.2 and 30/1.2, respectively. Current generators IB1-IB4 are equal to 10 µA, while IB5-IB6 are 20 µA. For an assumed input capacitance of 1 pF the required compensation capacitor was 0.5 pF. Table 2.7 reports the main parameters.

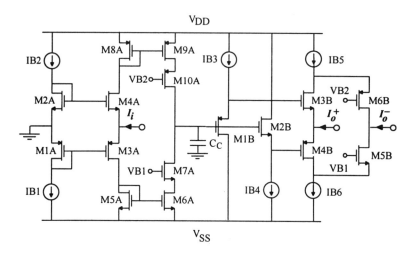

Fig. 2.23. *Detailed schematic of the first VFCOA design*

Table 2.7
Main performance of the first VFCOA design

Parameter	Value	
Technology	1.2 µm	
V_{DD}-V_{SS}	5 V	
DC Power	0.42 mW	
Gain	62 dB	
GBW	30 MHz	
M_ϕ (C_L=1pF)	53 °	
$T_{settl.}$(0.1%)	35 ns	
THD	-70 dB	-60 dB
(i_{out})	(5mA)	(10mA)

Other performance details can be gained from Fig. 2.24 to 2.26. Fig. 2.24 illustrates the bode plot of the loop gain (module and phase). The response to a ±5-µA square wave input with the amplifier in unity-gain configuration is shown in Fig. 2.25. Slew-Rate is about 0.4 µA/ns. Finally, a set of closed loop transfer functions for a fixed value of R_2 (10 kΩ) and R_1 variable is reported in Fig. 2.26. The closed-loop bandwidth is approximately constant. However for high closed-loop gain, when R_1 is comparable to the non-inverting output resistance (r_{o1} in equation (1.39)) the bandwidth starts to depend on gain and the transfer function asymptotically tends to a constant GBW behavior.

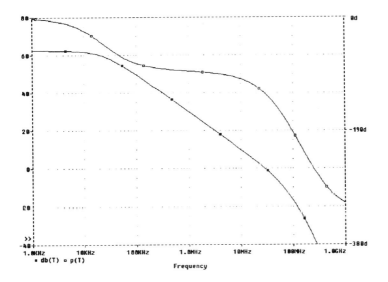

Fig. 2.24. Bode plot of the loop gain

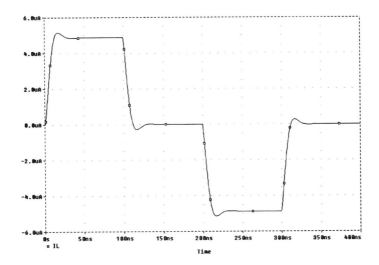

Fig. 2.25. *Time response of the amplifier connected in unity-gain configuration to a square input of ±5 μA.*

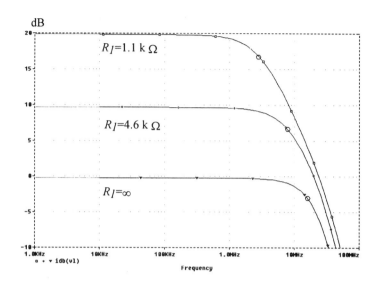

Fig. 2.26. Closed loop gain for $R_2 = 10\ k\Omega$ and variable R_1

B. Second design

We conclude this section by considering the implementation of the class AB VFCOA proposed by Kaulberg [34]. The circuit is shown in Fig. 2.27 and its made up of two class AB CCII-s. The second CCII has a cross-coupled output section to provide inversion in the output current flow. All transistors have a channel length of 4.8 μm, while nMOS and pMOS transistors have a width of 122 μm and 403 μm, respectively.

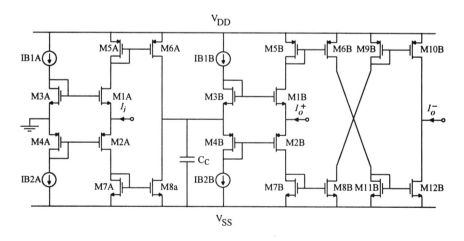

Fig.2.27. Simple class AB VFCOA

For the sake of simplicity we have drawn simple current mirrors in the schematic. But, in the actual design cascode current mirrors were adopted. This choice improves accuracy but limits amplifier current swing. Despite its class AB architecture, only a poor drive capability is achieved (hence the reason for including this circuit in the present chapter). Indeed, with a quiescent current in the output branch (M10B, M12B) of about 270 µA, the measured output swing is of ±700 µA. Unfortunately, we could not obtain information about the harmonic distortion. We believe, however, that it should be low thanks to the cascode mirrors adopted (which ensure equal operating conditions for the transistors in the mirrors) and the low ratio between the maximum output current and the quiescent current.

The main measured performance is summarized in Table 2.8. A constant bandwidth of about 1 MHz, for closed-loop gains ranging from 0 to 30 dB, is observed. In addition, high SR performance was found to be limited only by the amplifier bandwidth.

Table 2.8.
Main performance of the second VFCOA design

Parameter	Value
Technology	2.4 µm
V_{DD}-V_{SS}	5 V
Gain	72 dB
GBW	3 MHz
M_ϕ	60°
$T_{settl.}$(0.1%)	35 ns
Signal Range	± 700 µA

As a final remark, it is important to realize that this amplifier presents a conceptually different output stage from the previously discussed ones. Indeed, to achieve class AB operation, the current following action is essentially achieved through the use of current mirrors. This is one example of a larger class of output stages (also termed *translinear* COSs) that allow high-drive requirements to be satisfied. The drawback of translinear COSs is that they operate outside the feedback loop so that their performance is not enhanced by the negative feedback of the overall amplifier. Since this limitation principally affects linearity, suitable solutions have to be adopted to reduce harmonic distortion while preventing current swing. This topic will be developed in the next chapter.

2.4 CURRENT COMPARATORS

A current comparator is a useful basic building block for many circuits and applications such as Schmitt triggers, A/D converters, oscillators, current to frequency converters, neural networks, etc. [35-58]. A current comparator detects the difference between two (or more) input currents providing a two-level (*low* or *high*) output current. Since current comparators interface digital circuits which are often voltage-mode blocks, voltage output is usually provided, as shown in Fig. 2.28.

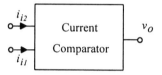

Fig. 2.28. Current comparator symbol

The Current Comparator can be regarded as a simpler version of an uncompensated low-drive COA. In this case, however, only one output terminal is needed. Nevertheless, design guidelines developed in the previous sections are not sufficient for high-performance current comparators, because optimized time response and offset compensation are also frequently required.

The original CMOS current comparator, i.e. the current-mirror comparator, is shown in Fig. 2.29 [59]. It is composed of a transresistance amplifier followed by an inverter stage.

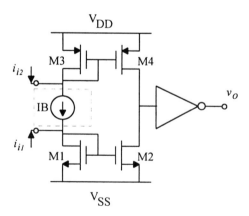

Fig. 2.29. Original current comparator

The transresistance amplifier is based on two complementary simple current mirrors that can be improved by using cascoded structures. Current IB was not present in the original version. The main limitation of this circuit lies in the absence of an appropriate input branch. As a result, parameters such as input resistance and input bias voltage, which greatly influence comparator performance, are not controlled unless an additional bias current is used (represented within the dashed box). Moreover, speed is reduced because one of the two output transistors (M2-M4) is in the triode region before comparison due to the high impedance of the output node [59].

As mentioned, fast response and accuracy are important requirements in comparators, so in this section design strategies will be developed for the optimization of these parameters. These techniques require clocked switches. However, this does not give rise to a real limitation, since clock signals are usually required in applications using comparators.

2.4.1 High-Speed Approaches

The main limitation to the time response in most current comparators comes from the output branch which is unbalanced at the beginning of the comparison phase. To achieve a fast time response the current comparator using a nonlinear positive feedback, shown in Fig. 2.30, was proposed [50], [51].

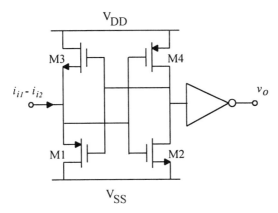

Fig. 2.30. Current comparator based on positive feedback

The circuit in Fig. 2.30 also has unbalanced output before comparison, but the positive feedback partially overcomes this drawback as it increases the input overdrive. Unfortunately, positive feedback applied at the input intrinsically leads to lower sensitivity which, in turn, means a low speed with a low input level. This approach also has other drawbacks which reduce the comparator performance. The input resistance is heavily dependent on the input signal and no control exists on the bias current, the effect of which can greatly increase power dissipation, especially in worst-case process and temperature conditions. Moreover, speed and sensitivity are drastically reduced if input signal generators with a relatively low internal resistance are used.

An alternative solution for increasing speed in current comparators adopts the pre-biasing technique [60]. Pre-biasing means properly setting the operating point before comparison to put the circuit in the best working condition at the beginning of the transient response. The technique is easily implemented in the current-mirror comparator using an inverter and a switch which shorts its input and output before comparation, as shown in Fig. 2.31.

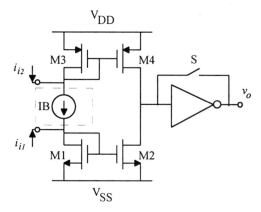

Fig. 2.31. Current comparator with pre-biasing

When the switch is on, the output of both the comparator and the inverter are set to a voltage nearly equal to $(V_{DD}+V_{SS})/2$. Pre-biasing was also adopted in the switched-current comparator reported in [61], but the solution in Fig. 2.31 uses only one switch and is more efficient because it initially sets the output at half rather than the whole supply voltage.

A comparison between the time responses of the original comparator (Fig. 2.29), the positive-feedback comparator (Fig. 2.30), and the original comparator with pre-biasing (Fig. 2.31) is shown in Fig. 2.32. The circuits were simulated using SPICE and the model parameters of a 2-μm CMOS process. The bias current shown in the dashed block of Figs. 2.29 and 2.31 has been also included in the simulation.

Curve 1 in Fig. 2.32 is the input step current (i.e., $i_{i1}-i_{i2}$), curves 2, 3 and 4 refer to the circuits in Fig. 1a, 1b and 1c, respectively. The pre-biased comparator has better rise and fall times than those of the other comparators, but needs a time interval for pre-biasing. In the actual simulation, the pre-bias time is 40 ns, but, as shown in curve 4, it can be reduced to less than 15 ns. Hence, if we also include pre-bias time, the overall switching time of the pre-biased comparator is similar to that of the positive-feedback comparator, but with better sensitivity.

The sensitivity of the comparator in Fig. 2.31 can be further improved by adopting cascode or Wilson current mirrors to greatly increase the transresistance gain of the circuit. However, better speed performance can also be obtained by using the double folded-cascode configuration in Fig. 2.33 [62], which is based on common-gate stages instead of current-mirror stages. Indeed, the common-gate stage exhibits a better frequency response than that of a current mirror, even a simple one.

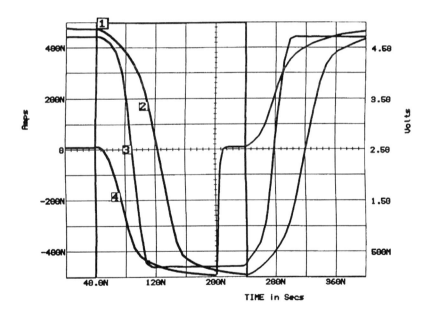

Fig. 2.32. Time responses:
1) input step current,
2) original current comparator output,
3) positive-feedback current comparator output and
4) current comparator with pre-biasing output

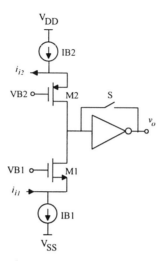

Fig. 2.33. Double folded-cascode current comparator

2.4.2 Design Control Considerations

A key aspect in the design of high-performance comparators is the control of parameters such as input resistance, input bias voltage, and the input and output bias current. This is because:

• a well-defined low input resistance is required to achieve high sensitivity. In fact, an input resistance which is dependent on the input level, like the circuits in Fig. 1 without IB, reduces sensitivity for low input currents due to the current partition caused by the finite internal impedance of real current generators.
• a well-defined input bias voltage is required in order to allow input generators to be properly biased (remember that current generators are usually made up of transistors in the saturation region).
• well-controlled input and output bias currents are required in order to control time response and power dissipation.

Current comparators based on positive feedback are intrinsically uncontrolled, while the comparators in Fig. 2.29 and 2.31 can easily be controlled by feeding the current mirrors with a bias current as shown in the dashed blocks. Of course, the double folded-cascode comparator in Fig. 2.33 is inherently controlled.

Current comparators based either on current mirrors or on the double folded-cascode topology can be further improved by adopting an appropriate input stage. At present, the best solution seems to be the class AB source-coupled stage which was already discussed and shown in Fig. 2.13c. It is worth noting that when the circuit in Fig. 2.13c is applied to current-mirror comparators it provides a true class AB comparator given that the current in the circuit will be as high as the input current.

2.4.3 Offset Compensation

Offset is one of the most critical parameters which limits comparator performance. Hence, offset compensation is a target worth pursuing in high-performance comparators. The first offset-compensated current comparator was presented in [63], but this approach is rather complicated and cannot easily be extended to other topologies. Subsequently, two more general offset-compensation techniques were proposed which can easily be applied to different comparator architectures. They use an additional compensation circuit embedded in parallel or in series in the uncompensated comparator. The following section discusses these two compensation techniques together with the circuit arrangements needed to compensate the charge-injection error as well.

A. First approach: parallel-connected compensation circuit

A first solution for offset compensation is schematically illustrated in Fig. 2.34. It uses a compensation circuit based on a hold capacitor, C_H, a switch, SA, and a transconductance stage whose output is connected in parallel to the uncompensated comparator.

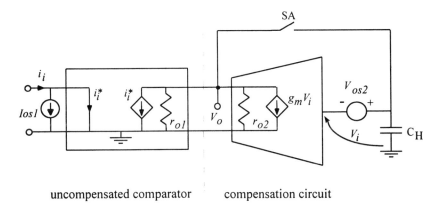

Fig. 2.34. Block diagram of a current comparator using the parallel-connected offset compensation circuit

In order to discuss circuit behavior, the uncompensated comparator has been represented by a unity-gain current amplifier with an output resistance, r_{o1}, and an input offset current, I_{os1}. The transconductance stage in the compensation circuit is characterized by a transconductance gain, g_m, an output resistance, r_{o2}, and an input offset voltage, V_{os2}. Assuming as initial conditions switch SA being open and capacitor C_H being discharged, the output offset voltage, V_{os}, of the overall circuit becomes

$$V_{os} = r_o I_{os1} + g_m r_o V_{os2} \tag{2.47}$$

where r_o is the equivalent output resistance

$$r_o = r_{o1} \| r_{o2} \tag{2.48}$$

The equivalent offset current to the comparator input is

$$I_{os} = I_{os1} + g_m V_{os2} \tag{2.49}$$

When switch SA is closed, the transconductance stage is connected in a unity-gain configuration and the output offset voltage, V_{osc}, becomes

$$V_{osc} = \frac{V_{os}}{1 + g_m r_o} \cong \frac{I_{os1}}{g_m} + V_{os2} \tag{2.50}$$

where term $g_m r_o$ is the loop-gain of the feedback transconductance stage. When switch SA turns off, voltage V_{osc} is stored in capacitor C_H, thus maintaining its value in the output node. The equivalent offset current, I_{osc}, to the comparator input is now given by

$$I_{osc} = \frac{V_{osc}}{r_o} \cong \frac{I_{os}}{g_m r_o} \tag{2.51}$$

Consequently, a very low input offset current is achieved despite the additional offset component due to the compensation circuit,. Actually, the overall offset current before compensation I_{os} is reduced by one stage gain.

Of course, the input signal has to be fed after switch SA is opened.

The application of such a technique to the current-mirror comparator is shown in Fig. 2.35.

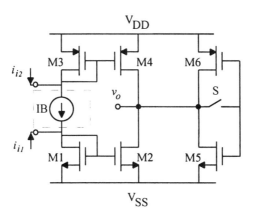

Fig. 2.35. The current-mirror comparator using the parallel-connected offset compensation circuit

To preserve the simplicity of the original comparator, an inverter is used as transconductance stage. Its input capacitance provides the storage element, C_H. Of course, cascode transconductance stages must be used with cascode-mirror or folded-cascode comparators.

B. Charge-injection compensation

The offset compensation topology in Fig. 2.34 is affected by a charge-injection error. When switch SA is opened a portion ΔQ of its channel charge is pushed into C_H, giving rise to an uncompensated residual offset. The charge-injection offset can be included in the input offset current by simply modifying equation (2.51) into

$$I_{osc} \cong \frac{I_{os}}{g_m r_o} + g_m \Delta V_{ck} \qquad (2.52)$$

where ΔV_{ck}, which is given by

$$\Delta V_{ck} = \frac{\Delta Q}{C_H} \qquad (2.53)$$

is the offset voltage at the input of the transconductance stage due to the charge injection. The charge ΔQ injected into C_H depends on the total amount of the switch channel charge, the shape of the clock edges, and the ratio between the capacitances at the switch terminals [64-65]. In order to reduce it, common remedies are the use of a CMOS switch, a larger capacitance C_H, and/or the use of a dummy switch.

A more efficient charge-injection compensation technique is to adopt a differential-input transconductance stage as shown in Fig. 2.36 [66]. Since switches SA1 and SA2 have equal voltages at their terminals, and assuming a clock phase with very sharp edges, the charges injected into C_{H1} and C_{H2} are approximately equal, regardless of the impedance at the switch terminals. Therefore, if capacitors C_{H1} and C_{H2} are equal, the injected charges will not affect the comparator output, since they give rise to a common mode signal which is rejected by the differential transconductance stage.

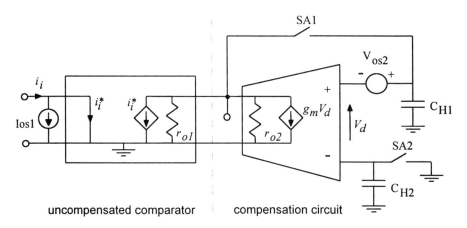

Fig. 2.36. Block diagram of a current comparator using the parallel-connected offset and charge-injection compensation circuit

A circuit implementation of the block scheme in Fig. 2.36 for the current-mirror comparator is illustrated in Fig. 2.37.

To validate the charge-injection compensation, the circuit was simulated by SPICE using minimum area CMOS switches. The simulation results are illustrated in Fig. 2.38. They were obtained without an input signal and considering only an input offset current of 0.4 µA and the charge-injection error.

In Fig. 2.38, curve 1 refers to the circuit without charge-injection compensation (i.e., without C_{H2} and SA2, and with the gate of M6 to ground), while curve 2 refers to the circuit with charge-injection compensation. In the first 20 ns the circuit is uncompensated and shows an output offset voltage which is forced by the input offset current. During the time interval from 20 ns to 60 ns, the switches are closed and the offset is compensated. After 60 ns, the switches are open and the charge-injection error occurs. The simulation shows that this error is greatly reduced by the compensation circuit.

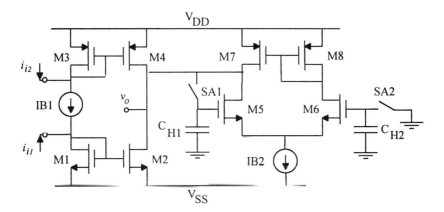

Fig. 2.37. The current-mirror comparator using the parallel-connected offset and charge-injection compensation circuit

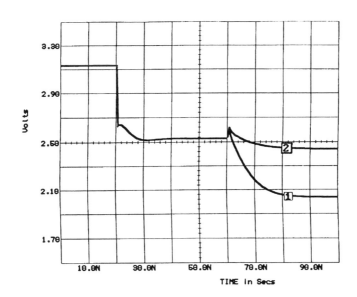

*Fig. 2.38 Charge-injection simulation for the circuit in Fig. 2.37
1) without compensation and 2) with compensation*

C. Second approach: series-connected compensation circuit

Both the solutions in Figs. 2.34 and 2.36 can easily be applied to most current comparators, since the compensation circuit is connected in parallel to the output branch. This means that no constraints need be imposed on the frequency response of the uncompensated comparator, since stability is guaranteed so long as the compensation circuit is stable. Unfortunately, with these solutions the overall comparator gain, and hence sensitivity, both decrease due to compensation circuit reducing the output resistance. To overcome this drawback, compensation circuits whose output resistance is higher than that of the uncompensated comparator would be needed. However, such a requirement is difficult to satisfy especially when cascode comparators are used.

An alternative but less common approach to offset compensation not affecting the comparator gain is shown in Fig. 2.39. It was first introduced in [67] and then optimized in [62] for both offset and charge-injection compensation.

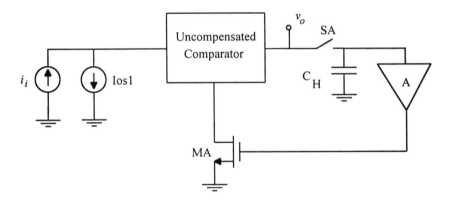

Fig. 2.39. Block diagram of a current comparator using the series-connected offset compensation circuit

This technique is based on a feedback loop which includes a switch, SA, a voltage gain stage, block A, and a current generator transistor, MA, which belongs to the output branch of the uncompensated comparator. This approach is quite similar to the parallel solution in Fig. 2.34, but in this case the transconductance element, which is performed by transistor MA, is in series with the output branch. Thus, unlike for the parallel solution, the output impedance of the original comparator can now be preserved.

As far as offset performance is concerned, analysis of the circuit in Fig. 2.34 can easily be extended to the series-connected compensation circuit provided that the transconductance g_m in (2.49) and (2.51) is replaced by the transconductance $A_o g_{mA}$ (A_o is the voltage gain of block A). Therefore, the uncompensated and compensated equivalent offset currents to the comparator input become, respectively

$$I_{os} = I_{os1} + A_o g_{mA} V_{os2} \tag{2.54}$$

$$I_{osc} \cong \frac{I_{os}}{A_o g_{mA} r_o} \tag{2.55}$$

where V_{os2} is the input offset voltage of block A, and r_o is the equivalent output resistance of the comparator. Of course, the output branch of the uncompensated comparator in the series solution affects the loop performance when switch SA is closed.

Block A can be implemented with a source follower since the gain $g_{mA} r_o$ is usually high enough to provide accurate offset compensation. Presently, the best implementation of the solution in Fig. 2.39 is the one based on the double folded-cascode current comparator. For example, by using the circuit in Fig. 2.33 and substituting current generator IB2 with transistor MA, the compensated comparator shown in Fig. 2.40 is achieved.

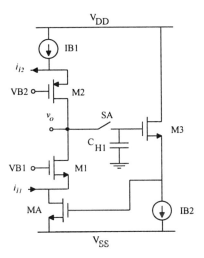

Fig. 2.40. *The double folded-cascode current comparator using the series-connected offset compensation circuit*

The natural evolution of the series-connected compensation circuit to achieve charge-injection compensation is shown in Fig. 2.41. Similarly to the solution given in Fig. 2.36, block A in Fig. 2.41 has been replaced by a differential stage. A circuit implementation based on the double folded-cascode current comparator is shown in Fig. 2.42.

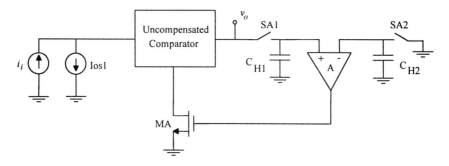

Fig. 2.41. Block diagram of a current comparator using the series-connected offset and charge injection compensation circuit

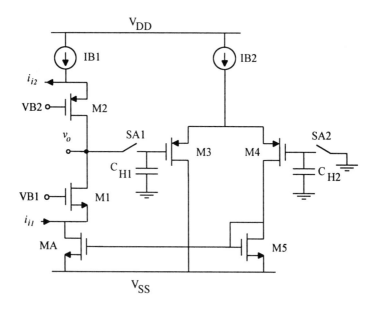

Fig. 2.42. The double folded-cascode current comparator using the series-connected offset and charge injection compensation circuit

2.4.4 Design Examples

In this section we give two examples of current comparators which adopts most of the techniques developed. We start with a simple design and conclude with a high-performance fully differential comparator.

A. Offset-compensated current comparator

The first example of current comparator is shown in Fig. 2.43 [62]. It is made up of a transresistance amplifier (transistors M1-M19) and series connected offset-compensation circuit (transistors M20-M21, switch SA and hold capacitor C_H).

Assuming switch SA to be closed, the current in the input branch is imposed on the output branches (transistors M12-M15 and M16-M19) thanks to the cascode current mirrors. They provide a high output resistance and accurate current mirroring action.

With switch SA open, transistors M6-M19 become a high-gain transresistance amplifier. The output node will go high or low according to the sign of the input signal.

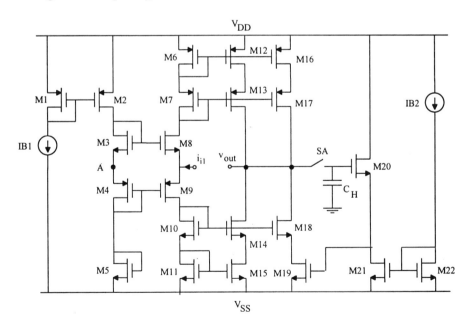

Fig. 2.43. *Schematic of the offset-compensated current comparator*

The offset-compensation mechanism is easily understood: at the beginning of the comparation phase, switch SA is turned off and the output voltage is frozen on capacitor C_H, thus the input offset is kept unchanged.

The circuit is designed in a 2-μm process. Transistor aspect ratios are shown in Table 2.9. Reference currents IB1 and IB2 are set to 1 μA and 12 μA, respectively and a 5-V power supply is used. Finally, the hold capacitor is set to 1 pF.

Table 2.9.

Transistor aspect ratios of the circuit in Fig. 2.43

Transistor	W/L (μm/μm)
M1, M2	10/2
M3, M8	6/2
M4, M9	12/2
M5	2/30
M6, M7, M12, M13 M16, M17	9/2
M10, M11, M14, M15 M18, M19	3/2
M20	2/4
M21, M22	20/4

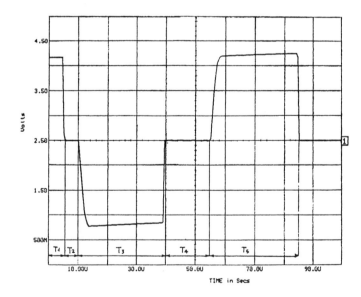

Fig. 2.44. Output response with an input current of ±30 nA forcing an input offset current of 50 nA

An output time response to an input step current of ±30 nA is shown in Fig. 2.44. An input offset current of 50 nA is also forced onto the input node. This offset is responsible for an output offset voltage of 1.5 V before compensation, as shown during time interval T1. During time intervals T2 and T4 switch SA is closed and the output voltage is set to around $V_{DD}/2$. During time intervals T3 and T5 a negative and a positive 30-nA input step is applied causing the output voltage to go low or high, respectively.

B. A Fully-differential comparator

A fully-differential topology has the inherent advantages of accuracy and power supply noise rejection. Moreover, the availability of a differential input could be very useful in applications where the input is a floating source. In this section, a high-performance fully-differential current comparator is presented which adopts most of the design arrangements discussed in the previous sections [68].

The comparator schematic is shown in Fig. 2.45. It is based on the double folded-cascode structure and includes a compensation circuit which provides offset and charge-injection compensation as well as common-mode output voltage control.

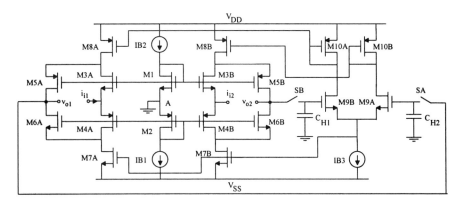

Fig. 2.45. Schematic of the fully-differential current comparator

The uncompensated comparator is made up of transistors M3-M6 and current source transistors M7, M8. Diode-connected transistors M1 and M2, and current sources IB1 and IB2, set the input bias current and the input bias voltage. The compensation circuit is provided by the differential stage, M9-M10, the current generator, IB3, the storage capacitors, C_{H1} and C_{H2}, and the switches, SA and SB. Thanks to the perfect symmetry of the circuit, the diode-connected transistors M10A, M10B can set the bias current in M8A and M8B to IB3/2. In addition, the gate-source voltage of M9 together with the gate-source voltage of M7 provide the output bias voltage.

When switches SA and SB are closed, the uncompensated comparator and the compensation circuit are connected through two different loops, one for the differential signal, the other for the common-mode signal. When switches SA and SB are opened, the two loops are disconnected and the common-mode output level and the output offset voltage are both frozen in the hold capacitors. As far as the differential loop is concerned, the circuit can be represented with a block diagram which is a differential version of the one in Fig. 2.42. Hence, the charge injection is intrinsically compensated since it appears as a common mode signal. The input offset current after compensation is given by (2.55) so long as transconductance g_{mA} is substituted with that provided by transistors M8 and gain A_O is defined as the differential gain of the compensation circuit (i.e., g_{m9}/g_{m10}). Thus, the input offset current is given by

$$I_{osc} \cong \frac{g_{m10}}{g_{m9}} \frac{I_{os}}{g_{m8} r_o} \qquad (2.56)$$

The circuit in Fig. 2.45 is simulated with a 2-μm CMOS process. The transistor aspect ratios and the other main design parameters are shown in Table 2.10.

Table 2.10.
Electrical parameters of the circuit in Fig. 2.45

Parameter	Value
M1, M3A, M3B, M6A, M6B	3/2 (μm/μm)
M2, M4A, M4B, M5A, M5B, M8A, M8B, M10A, M10B	8/2 (μm/μm)
M7A, M7B, M9A, M9B	2/4 (μm/μm)
C_{H1}, C_{H2}	1 pF
IB1, IB2	1 μA
IB3	4 μA
V_{DD}-V_{SS}	5 V

By setting the power supply to 5 V, the overall power dissipation come to around 45 μW. A sensitivity simulation with a slowly varying triangular input current is shown in Fig. 2.46, where curve 1 is the input current and curve 2 is the output voltage. A better than 20 nA sensitivity is achieved.

A transient simulation with an input step current from -0.5 µA to 0.5 µA is illustrated in Fig. 2.47. The switching time from 2.5 V to 3.5 V is shorter than 30 ns, while that from 2.5 V to 1.5 V is shorter than 20 ns.

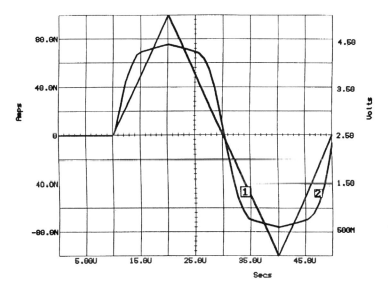

Fig. 2.46. Sensistivity simulation of the comparator in Fig. 2.45:
1) input current, 2) output voltage

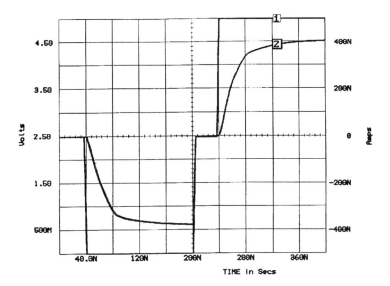

Fig. 2.47. Output response of the comparator in Fig. 2.45:
1) input step, 2) output voltage

REFERENCES

[1] G. C. Temes, W. H. Ki, "Fast CMOS Current Gain Amplifier and Buffer Stage," *Electronics Letters*, Vol.23, pp. 696-697, 1987.

[2] Z. Wang, W. Guggenbühl, "Adjustable Bidirectional MOS Current Mirror/Amplifier," *Electronics Letters*, Vol.25, No. 10, pp. 673-675, May 1989.

[3] E. A. M. Klumperink, H. J. Janssen, "Complementary CMOS Current Gain Cell," *Electronics Letters*, Vol.27, pp. 38-40, 1991.

[4] Z. Wang, "Wideband Class AB (Push-Pull) Current Amplifier in CMOS Technology," *Electronics Letters*, Vol.26, No. 3, pp. 543-545, Apr. 1990.

[5] A. Sedra, K. Smith, *Microelectronic Circuits*, CBS College Publishing, 1987.

[6] A. Sedra, K. Smith, "A Second-Generation Current Conveyor and Its Applications", *IEEE Trans. on Circuit Theory*, CT-17, pp. 132-133, Feb. 1970.

[7] A. Sedra, G. Roberts, F. Gohh, "The Current Conveyor: History, Progress and New Results," *IEE Proc. Part G*, Vol.137, No.2, pp.78-87, Apr. 1990.

[8] P. Aronhime, "Transfer-Function Synthesis Using a Current Conveyor," *IEEE Trans. on Circuits and Systems,* pp.312-313, Mar. 1974.

[9] A. Fabre, M. Alami, "Insensitive Current-Mode Bandpass Implementations-Based Nonideal Gyrators," *IEEE Trans. on Circuits and Systems - part. I*, Vol. 39, No. 2, pp. 152-155, Feb. 1992.

[10] G. Roberts, A. Sedra, "A General Class of Current Amplifier-Based Biquadratic Filter Circuits," *IEEE Trans. on Circuits and Systems -part. I*, Vol. 39, No. 4, pp. 257-263, Apr. 1992.

[11] C. Chang, "Universal Active Current Filter with Single Input and Three Outputs Using CCIIs," *Electronics Letters*, Vol. 29, No. 22, pp. 1932-1933, Oct. 1993.

[12] A. Fabre, F. Dayoub, L. Duruisseau, M. Kamoun, "High Input Impedance Insensitive Second-Order Filters Implemented from Current Conveyors," *IEEE Trans. on Circuits and Systems - part I*, Vol. 41, No. 12, pp. 918-921, Dec. 1994.

[13] R. Nandi, "Insensitive Current Mode Realization of Third-Order Butterworth Characteristics Using Current Conveyors," *IEEE Trans. on Circuits and Systems - part I*, Vol. 41, No. 12, pp. 925-927, Dec. 1994.

[14] P. Mohan, "New Current-Mode Biquad on Friend-Deliyannis Active RC Biquad," *IEEE Trans. on Circuits and Systems - part II*, Vol. 42, No. 3, pp. 225-228, Mar. 1995.

[15] A. Fabre, M. Alami, "Universal Current Mode Biquad Implemented from Two Second Generation Current Conveyors," *IEEE Trans. on Circuits and Systems - part I*, Vol. 42, No. 7, pp. 383-385, July 1995.

[16] S. Liu, J. Chen, Y. Hwang, "New Current Mode Biquad Filters using Current Followers," *IEEE Trans. on Circuits and Systems - part I*, Vol. 42, No. 7, pp. 380-383, July 1995.

[17] A. Fabre, H. Amrani, O. Saaid, "Current-Mode Band-Pass Filters with Q-Magnification," *IEEE Trans. on Circuits and Systems - part II*, Vol. 43, No. 12, pp. 839-842, Dec. 1996.

[18] M. Abuelma'atti, A. Al-Ghumaiz, "Novel CCI-Based Single-Element-Controlled Oscillators Employing Grounded Resistors and Capacitors," *IEEE Trans. on Circuits and Systems - part I*, Vol. 43, No. 2, pp. 153-155, Feb. 1996.

[19] H. Elwan, A. Soliman, "A Novel CMOS Current Conveyor Realization with an Electronically Tunable Current Mode Filter Suitable for VLSI," *IEEE Trans. on Circuits and Systems - part II*, Vol. 43, No. 9, pp. 663-670, Sept. 1996.

[20] A. Soliman, "Generation of Current Conveyor-Based All-Pass Filters from Op Amp-Based Circuits," *IEEE Trans. on Circuits and Systems - part II*, Vol. 44, No. 4, pp. 324-330, Apr. 1997.

[21] K. Smith, A. Sedra, "Realisation of the Chua Family of New Nonlinear Network Elements Using the Current Conveyor," *IEEE Trans. on Circ. Theory*, pp.137-139, Feb. 1970.

[22] S. Liu, D. Wu, H. Tsao, J. Wu, J. Tsay, " Nonlinear Circuit Applications with Current Conveyors," *IEE Proc. Part G*. Vol.140, No.1, pp.1-6, Feb. 1993.

[23] G. Di Cataldo, G. Palumbo, S. Pennisi, "A Schmitt Trigger by Means Of a CCII+," *Int. J. of Circuit Theory and Applications*, Vol. 23, No.2, pp. 161-165, Mar. 1995.

[24] E. Bruun, "Analysis of the Noise Characteristics of CMOS Current Conveyors," *Int. J. Analog Integrated Circuits and Signal Processing*, No.12, pp. 71-78, 1997

[25] W. Surakampontorn, V. Riewruja, K. Kumwachara, K. Dejhan, "Accurate CMOS-based Current Conveyors," *IEEE Trans. on Instrumentation and Measurement*, Vol.40, No.4, pp.699-702, Aug. 1991.

[26] G. Palmisano, G Palumbo "A Simple CMOS CCII+," *Int. J. of Circ. Theory and Applications*, Vol. 23, no.6, pp. 599-603, Nov. 1995.

[27] Th. Laopoulos, S. Siskos, M. Bafleur, Gh. Givelin, "CMOS Current Conveyor," *Electronics Letters*, Vol.28, No.24, pp.2261-2262, Nov. 1992.

[28] G. Palmisano, G. Palumbo, S. Pennisi, "Design Strategies for Class A CMOS CCIIs", in print on *Int. J. of Analog Integrated Circuits and Signal processing*.

[29] P. Crawley, G. Roberts, "High-Swing MOS Current Mirror with Arbitrarily High Output Resistance," *Electronics Letters*, Vol. 28, No. 4, pp. 361-363, Feb. 1992.

[30] A. F. Arbel and L. Golminz, "Output Stage for Current-Mode Feedback Amplifiers, Theory and Applications," *Int. J. Analog Integrated Circuits and Signal Processing*, Vol. 2,3, pp. 243-255, 1992.

[31] E. Bruun, "A High-Speed CMOS Current Opamp for Very Low Supply Voltage Operation," *Proc. IEEE ISCAS'94*, London, 1994.

[32] E. Bruun, "Bandwidth Optimization of a Low Power, High Speed CMOS Current Op Amp," *Int. J. Analog Integrated Circuits and Signal Processing*, No.7, pp. 11-19, 1995

[33] E. Abou-Allam, E. El-Masry, "A 200 MHz Steered Current Operational Amplifier in 1.2-mm CMOS Technology," *IEEE J. of Solid-State Circuits*, Vol.32, No.2, pp. 245-249, Feb. 1997.

[34] T. Kaulberg, "A CMOS Current-Mode Operational Amplifier," *IEEE J. of Solid-State Circuits,* Vol.28, No.7, pp.849-852, July 1993.

[35] Z. Wang, W. Guggenbthl, "Novel CMOS Current Schmitt Trigger," *Electronics Letters*, Vol.24, No.24, pp.1514-1516, November 1988.

[36] Z. Wang, W. Guggenbthl, "CMOS Current Schmitt Trigger with Fully Adjustable Hysteresis," *Electronics Letters*, Vol.25, No.6, pp.397-398, March 1989.

[37] G. Di Cataldo, G. Palumbo "New CMOS Current Schmitt Trigger," *Proc. IEEE ISCAS'92*, May 1992.

[38] G. Di Cataldo, G. Palmisano, G. Palumbo "Low Area Accurate CMOS Current Schmitt Trigger," *Proc. ECCTD'93*, Sept. 1993.

[39] P. Crolla, "A Fast Latching Current Comparator for 12-Bit A/D Applications," *IEEE J. of Solid-State Circuits*, Vol. SC-17, No.6, pp.1088-1093, Dec. 1982.

[40] J. Robert, P. Deval, G. Wegmann, "Novel CMOS Pipelined A/D Convertor Architecture Using Current Mirrors," *Electronics Letters*, Vol.25, No.11, pp.691-692, May 1989.

[41] D. Nairn, C. Salama, "Current-Mode Algorithmic Analog-to-Digital Converters," *IEEE J. of Solid-State Circuits*, Vol.25, No.4, pp.997-1004, Aug. 1990.

[42] Chu P. Chong, "A Technique for Improving the Accuracy and the Speed of CMOS Current-Cell DAC," *IEEE Trans. on Circuits and Systems*, Vol.37, No.10, pp.1325-1327, Oct. 1990.

[43] D. Nairn, C. Salama "A Ratio-Independent Algorithmic Analog-to-Digital Converter Combining Current Mode and Dynamic Techniques" *IEEE Trans. on Circuits and Systems*, Vol.37, No.10, pp.319-325, March 1990.

[44] Z. Wang, "Design Methodology of CMOS Algorithmic Current A/D Converters in View of Transistor Mismatches," *IEEE Trans. on Circuits and Systems*, Vol.38, No.6, pp.660-667, June 1991.

[45] Seong-Won Kim, Soo-Won Kim, "Current-Mode Cyclic ADC for Low Power and High Speed Applications," *Electronics Letters*, Vol.27, No.10, pp.818-820, May 1991.

[46] C. Wey, "Concurrent Error Detection in Current-Mode A/D Convertors," *Electronics Letters*, Vol.27, No.25, pp.2370-2372, Dec. 1991.

[47] W. Krenik, R. Hester, R. DeGroat, "Current-Mode Flash A/D Conversion Based on Current-Splitting Techniques," *Proc. IEEE ISCAS'92*, 1992.

[48] A. Cujec, C. Salama, D. Nairn, "An Optimized Bit Cell Design for a Pipelined Current-Mode Algorithmic A/D Converter," *Int. J. Analog Integrated Circuits and Signal Processing*, No.3, pp.137-141, 1993.

[49] C. Wey, S. Krishnan, S. Sahli, "Design of Concurrent Error Detectable Current-Mode A/D Converters for Real-Time Applications," *Int. J. Analog Integrated Circuits and Signal Processing*, No.4, pp.65-74, 1993.
[50] K. Wong, K. Chao, "Current -Mode Cyclic A/D Conversion Technique," *Electronics Letters*, Vol.29, N.3, pp.249-250, Feb. 1993.
[51] K. Fong, C. Salama, "Low-Power Current-Mode Algorithmic ADC," *Proc. IEEE ISCAS'94*, May 1994.
[52] A. Cable, R. Harjani, "A 6-Bit 50MHz Current-Subtracting Two Step Flash Converter," *Proc. IEEE ISCAS'94*, May 1994.
[53] L. Zhang, T. Sculley, T. Fiez, "A 12 Bit, 2V Current-Mode Pipelined A/D Converter Using a Digital CMOS Process," *Proc. IEEE ISCAS'94*, May 1994.
[54] M. Yamamoto, A. Kobayashi, Y. Horio, "Switched Current F/I and I/F Converters," *Proc. ECCTD-91*, Sept. 1991.
[55] K. Current, J. Current, "CMOS Current-Mode Circuits for Neural Network," *Proc. IEEE ISCAS' 90*, 1990.
[56] K. Current, "Algorithmic Analogue-to-Quaternary Convertor Circuit Using Current-Mode CMOS," *Electronics Letters*, Vol.28, No.12, pp.1111-1112, June 1992.
[57] H. Gustat, "Fast CMOS Multilevel Current Comparator," *Electronics Letters*, Vol.29, No.7, pp.592-593, Apr. 1993.
[58] K. Current, "Current-Mode CMOS Multiple-Valued Logic Circuits," *IEEE J. of Solid-State Circuits*, Vol.29, No.2, pp.95-107, Feb. 1994.
[59] D.Freitas, K. Current, "CMOS Current Comparator Circuit," *Electronics Letters*, Vol.19, No.17, pp.695-697, Aug. 1983.
[60] G. Palmisano, G. Palumbo, S. Pennisi, "A High-Accuracy, High-Speed CMOS Current Comparator," *Proc. IEEE ISCAS'94*, May 1994.
[61] J. Carreira, J. Franca, "High-Speed CMOS Current Comparators," *Proc. IEEE ISCAS'94*, May 1994.
[62] G. Palmisano, G. Palumbo, "Offset-Compensated Low Power Current Comparators", *Electronics Letters*, Vol.30, No.20, pp.1637-1639, Sept. 1994.
[63] C. Wu, C. Chen, M. Tsai, C. Cho, "A 0.5mA Offset-Free Current Comparator for High Precision Current-Mode Signal Processing," *Proc. IEEE ISCAS'91*, 1991.
[64] J. Shieh, M. Patil, B. Sheu, "Measurement and Analysis of Charge Injection in MOS Analog Switches," *IEEE J. of Solid-State Circuits*, Vol. SC-22, No.2, pp.277-281, Apr. 1987.
[65] G. Wegmann, E. Vittoz, F. Rahali, "Charge Injection in Analog MOS Switches," *IEEE J. of Solid-State Circuits*, Vol. SC-22, No.6, pp.1091-11097, Dec. 1987.
[66] F. Maloberti, G. Palmisano, G. Torelli, "A Novel Approach for High-Frequency Gain-Compensated Sample-and-Hold Circuits," *Proc. IEEE ISCAS'94*, May 1991.
[67] G. Di Cataldo, G. Palmisano, G. Palumbo, S. Pennisi, "An Accurate Offset-Compensated Current Comparator," *Proc. IEEE MIDWEST'94*, 1994.

[68] G. Palmisano, G. Palumbo, "High Performance CMOS Current Comparator Design," *IEEE Trans. on Circuits and Systems*, Vol.12, No.12, pp.785-790, Dec. 1996.

Chapter 3

HIGH-DRIVE CURRENT AMPLIFIERS

The current amplifiers described in the previous chapter can profitably be used for on-chip signal processing, but their poor drive capability makes them unsuitable for driving off-chip loads. In these cases, a current amplifier with a class AB output stage is mandatory. To this end, the high-drive current amplifier becomes the natural front-end block for current-mode ICs.

The design of driver stages in power amplifiers is an important but difficult task, especially for current-mode circuits. As we have already pointed out, a class AB current output stage which uses current mirrors to produce multiple outputs works outside the feedback loop. Therefore, the linearity of the system is conditioned by the linearity of the current mirrors involved. Consequently, classical solutions based on standard current mirrors can hardly be employed if high linearity performance is desired. At present, solutions preserving swing and providing reduced harmonic distortion need to be developed for the implementation of the output stage.

And yet, since the input stage of a high-gain amplifier has to deal with small signal amplitudes, the same topologies can be adopted both for low-drive and high-drive circuits. Thus, the present chapter shall mainly deal with class AB current output stages and their use in implementing high-performance current amplifiers. Due to the large-signal operating conditions, linearity in class AB current output stages is an important issue. It shall be discussed in detail and simplified equations derived for harmonic distortion over several solutions. We shall find that channel-length modulation and transistor mismatches are the main sources of distortion in this class of circuits. Moreover, we shall see that the non-linearity caused by channel-length modulation can be reduced by using circuit solutions, while mismatching between the transistors in a current mirror constitutes the fundamental limitation to linearity.

3.1 CLASS AB CURRENT OUTPUT STAGES

A critical, but often vital, section in most analog integrated circuits either for the current or voltage approach is the final power section that has to drive low resistive loads. In the voltage-mode approach, this part of the chip contains voltage power amplifiers. They must deliver high current and high voltage swing, and exhibit low harmonic distortion [1]-[5]. The latter is achieved to a great extent by a feedback connection. In fact, the output stage of a voltage amplifier works completely within the feedback loop. Current-mode power amplifiers use a current output stage (COS) as a final stage which is able to drive a grounded load with a bipolar current. The COS is the most critical block in the implementation of high-drive current amplifiers. Indeed, implementing such a stage leads to circuits being only partially included within the feedback loop. As a result, the output section becomes the main source of non-linearity.

Let us again consider the circuit in Fig. 2.28, which we have already identified as a first attempt to realize a current amplifier with a class AB output stage, and consider its use as a closed-loop unity-gain current buffer, schematically shown in Fig. 3.1. In brief, the current amplifier is made up of a transresistance amplifier, providing low input resistance and high gain, and a class AB COS, which has to provide high drive capability and high output resistance. More specifically, the transresistance amplifier includes an inner current amplifier (in the dashed block) and a push-pull voltage follower. The COS is implemented with two simple current mirrors, MA1-MA2 and MB1-MB2 and can be also regarded as a class AB current mirror. This is a simplification with respect to the circuit in Fig. 2.28 originally adopting two cross-coupled current mirrors to achieve current inversion. It is quite clear, however, that every current mirror used as a signal processing element introduces harmonic distortion. Thus, reducing the number of current mirrors needed is important when designing a linear COS. A minor drawback is the lack of current inversion causing a change in the sign of the closed loop gain. Fortunately, this does not constitute a problem in many applications.

High-drive Current Amplifiers

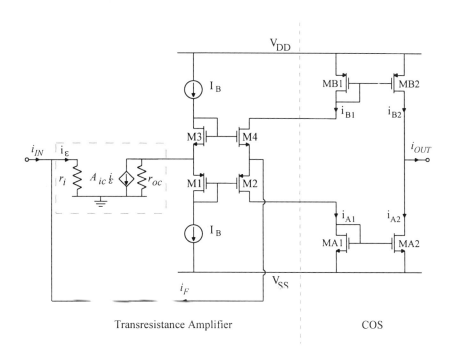

Fig. 3.1. Simplified schematic of a class AB current amplifier in unity-gain configuration

Thanks to the negative feedback around the transresistance amplifier, current i_F is a low-distorted feedback current whose value in the module is very close to the input current i_{IN}. The output current i_{OUT} is given by

$$i_{OUT} = i_{B2} - i_{A2} \tag{3.1}$$

If a COS with ideal unitary current mirrors is assumed (i.e., $i_{A2} = i_{A1}$ and $i_{B2} = i_{B1}$), it follows that

$$i_{OUT} = i_{B1} - i_{A1} = i_F \cong -i_{IN} \tag{3.2}$$

thus, an ideal inverting current buffer is achieved. Unfortunately, since real current mirrors operating under large signal conditions are affected by non-linearity, harmonic distortion is created in the output current. As already stated, the current mirrors operate outside the feedback loop, therefore the linearity of the system is limited by the linearity of the same. Since high performance COSs must exhibit high output impedance and accurate current transfer, simple current mirrors cannot be employed. Cascode mirrors, although they provide both high output resistance and linearity, have poor current swing [6]-[7] making them impractical for high-current applications. In conclusion, the use of current mirrors with cascoded output is mandatory in implementing a COS.

A COS has two main sources of non-ideality which cause deviation from the ideal DC transfer characteristic and affect linearity:

- the channel-length modulation error of the mirroring transistors (MA2 MB2 in Fig. 3.1);
- the mismatches between the transistors in the current mirror.

Channel-length modulation is caused by differences in the drain-source voltage of the transistors. It can be reduced by increasing their channel length. However, this means larger chip area and worse frequency response. At present, efficient solutions are respresented by circuit topologies which are suitably arranged to reduce the channel-length modulation effect and preserve the current swing [8]. To better understand the effect of channel-length modulation on linearity performance, let us consider the simple current mirror MA1-MA2 shown in Fig. 3.1, which is made up of two equal transistors, working in the saturation region. According to the relationship for the saturation region

$$i_D = \beta(v_{GS} - V_T)^2 (1 + \lambda v_{DS}) \qquad (3.3)$$

the output current of the n-type current mirror is given by

$$i_{A2} = \frac{1 + \lambda v_{DS_{A2}}}{1 + \lambda v_{GS_{A1}}} i_{A1} \qquad (3.4)$$

To have a linear relation, λ should be equal to zero or $v_{DS_{A2}}$ should be equal to $v_{GS_{A1}}$, which, in other words, means eliminating the channel-length modulation error on the mirroring transistors. Obviously, the second option is the only viable one from a circuit point of view, and solutions are presented towards this direction.

However, mismatches are due to transconductance-gain, β, and threshold-voltage tolerances in the mirroring transistors [9]-[11]. Mismatch errors similarly affect any topology and can only be reduced by careful layout design. They therefore place the greatest limitation on COS linearity.

To conclude, the design of high-performance class AB COSs translates into the development of current mirrors with high output swing, resistance and with reduced channel-length modulation effects. Several such topologies are illustrated in the following section.

3.1.1 Configurations

The simplest solution considered here is the COS based on regular cascoded mirrors shown in Fig. 3.2. The circuit exhibits high output resistance and swing, but any strategy allowing the drain voltage to track the gate voltage in the output transistors (i.e., MA2 and MB2) is lacking. Consequently, channel-length modulation, despite being less effective than in simple current mirrors, still limits linearity. We shall see that linearity is primarily affected by third-order harmonic distortion.

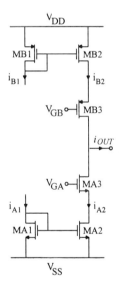

Fig. 3.2. Cascoded current output stage

A first solution which reduces channel-length modulation is using the cascoded mirror with dynamic matching shown in Fig. 3.3 [12]. It exhibits a reduced non-linearity by a factor of 2 thanks to the following action of common drain MA4 (MB4).

A better solution for reducing harmonic distortion due to channel-length modulation is achieved by implementing the COS using a current mirror with improved dynamic matching as shown in Fig. 3.4, proposed in [8]. However, close investigation reveals that while third-order harmonic distortion is minimized, a second-order harmonic distortion appears (normally negligible in other topologies) due to threshold voltage mismatches. Therefore, linearity may be still unsatisfactory for a high performance current amplifier. Note that a positive-feedback exists involving transistors MA6-MA8, but its loop-gain, $1/(g_{mA4}r_{dA5})$, is much lower than 1.

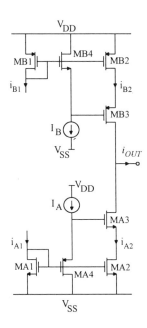

Fig. 3.3. Cascoded current output stage with dynamic matching

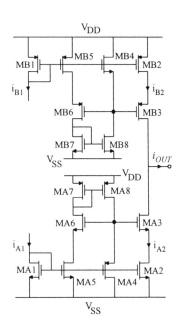

Fig. 3.4. Cascoded current output stage with improved dynamic matching

The last COS considered is shown in Fig. 3.5 [13]. It is made up of two complementary active-gain enhanced mirrors which base their performance on a principle quite similar to the gain-boosting technique [14]-[16]. The two current mirrors are composed of transistors MA1-MA3 and MB1-MB3 and two auxiliary voltage amplifiers, A1 and A2. Thanks to these, the drain voltages of MA1, MA2 and MB1, MB2 are almost equal even for large currents. Thus, a COS with high linearity performance is achieved. Moreover, the use of A1 and A2 also provides a very high output resistance given by

$$r_o \cong \left(g_{mA3}r_{dA2}r_{dA3}A_1\right) \| \left(g_{mB3}r_{dB2}r_{dB3}A_2\right) \qquad (3.5)$$

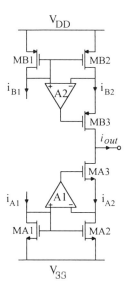

Fig. 3.5. Gain-boosted cascoded current output stage

3.2 HARMONIC DISTORTION DUE TO CHANNEL-LENGTH MODULATION

Let us evaluate the harmonic distortion due to the channel-length modulation in the previously discussed COSs. The most commonly used approaches [17]-[21] are not suitable for class AB amplifiers operating under large swing conditions, since the two half circuits work alternately. A useful approach for such circuits is suggested in [22] and has been rearranged for class AB COSs in Appendix 3.A. Another approach can be found in [23].

3.2.1 COS Based On Regular Cascoded Mirrors

Let us consider the COS with cascoded mirrors shown in Fig. 3.2. Without loss of generality, we only consider the *n*-type cascoded mirror (i.e., MA1-MA3) and assume transistors MA1-MA3 to be ideally matched and having the same transconductance gain (i.e., the same aspect ratio, W/L). For current mirror MA1-MA2, (3.4) still holds. But $V_{DS_{A2}}$ is now set by MA3 and V_{GA}. To guarantee accurate matching between MA1 and MA2 in quiescent conditions, voltage $V_{DS_{A2}}$ must be set equal to voltage $V_{GS_{A1}}$

($V_{DSA1} = V_{GSA1}$). The value of V_{GA} which provides such a condition is given by

$$V_{GA} = V_{DSA2} + V_{GSA3} \cong 2V_{TN} + 2\sqrt{\frac{I_Q}{\beta_N}} \qquad (3.6)$$

where I_Q is the quiescent current.

It should be noted that with single well technology, threshold voltages are affected by the body effect. This effect is a second-order source of non-linearity and will be neglected in the following analysis. From (3.4) it follows that

$$i_{A2} = \frac{1 + \lambda_N (V_{GA} - v_{GSA3})}{1 + \lambda_N v_{GSA1}} i_{A1} \cong \frac{1 + \lambda_N \left(2V_{TN} + 2\sqrt{\frac{I_Q}{\beta_N}} - V_{TN} - \sqrt{\frac{i_{A2}}{\beta_N}} \right)}{1 + \lambda_N v_{GSA1}} i_{A1}$$

(3.7a)

which can be approximated by

$$i_{A2} \cong \left[1 + \frac{2\lambda_N}{\sqrt{\beta_N}} \frac{\sqrt{I_Q} - \sqrt{i_{A1}}}{1 + \lambda_N \left(V_{TN} + \sqrt{\frac{i_{A1}}{\beta_N}} \right)} \right] i_{A1} \qquad (3.7b)$$

Taking the derivative of i_{A2} with respect to i_{A1} we obtain

$$\frac{\partial i_{A2}}{\partial i_{A1}} \cong 1 + \frac{2\lambda_N}{\sqrt{\beta_N}} \frac{\sqrt{I_Q} - \frac{3}{2}\sqrt{i_{A1}}}{1 + \lambda_N \left(V_{TN} + \sqrt{\frac{i_{A1}}{\beta_N}} \right)} \qquad (3.8a)$$

Since term $\lambda_N \left(V_{TN} + \sqrt{\frac{i_{A1}}{\beta_N}} \right)$ is usually much lower than 1, we get

$$\frac{\partial i_{A2}}{\partial i_{A1}} \cong 1 + \frac{2\lambda_N}{\sqrt{\beta_N}}\left(\sqrt{I_Q} - \frac{3}{2}\sqrt{i_{A1}}\right) \qquad (3.8b)$$

Following the same steps for the *p*-type current mirror MB1-MB3, and assuming the transconductance gain to be equal for both current mirrors (i.e., $\beta_N = \beta_P = \beta$), HD$_2$ and HD$_3$ are calculated from (3.A7) and (3.A8) of Appendix 3.A

$$HD_2 \cong \frac{\lambda_N - \lambda_P}{4}\sqrt{\frac{I_Q}{\beta}}\left(\frac{3}{2}\sqrt{\frac{I_M}{I_Q}} - 1\right) \qquad (3.9a)$$

$$HD_3 \cong \frac{\lambda_N + \lambda_P}{8}\sqrt{\frac{I_Q}{\beta}}\left(\sqrt{\frac{I_M}{I_Q}} - 1\right) \qquad (3.9b)$$

where I_M is the magnitude of the sinusoidal input current.

As expected from current mirrors with ideally matched transistors and equal transconductance gain, the even-order harmonic distortion is very low. In fact, it is proportional to the difference between the two channel-length modulation parameters. Therefore, third-order harmonic distortion becomes the dominant contribution.

Harmonic distortions HD$_2$ and HD$_3$ are dependent upon the relative magnitude of the input signal and λ_N and λ_P. They can be reduced by increasing the transconductance gain and/or the channel length of the transistors.

3.2.2 COS Based On Cascoded Mirrors With Dynamic Matching

The first circuit solution to reduce harmonic distortion in COS due to channel-length modulation was presented in [12] and is shown in Fig. 3.3. Thanks to the common drain MA4 (MB4), the gate voltage of the common-gate transistor MA3 (MB3) follows the gate voltage of MA1 (MB1), thus keeping V_{DSA2} (V_{DSB2}) constant. This improves linearity performance, since better dynamic matching between MA1 (MB1) and MA2 (MB2) is achieved.

Setting

$$V_{SG_{A4}} = V_{GS_{A3}} = -V_{TP_{A4}} + \sqrt{\frac{I_A}{\beta_{P_{A4}}}} = V_{TN} + \sqrt{\frac{I_Q}{\beta_N}} \qquad (3.10)$$

MA1 and MA2 are matched in quiescent conditions, and from (3.7a) we get

$$i_{A2} = \frac{1 + \lambda_N (v_{GS_{A1}} + v_{SG_{A4}} - v_{GS_{A3}})}{1 + \lambda_N v_{GS_{A1}}} i_{A1} \cong$$

$$\cong \left[1 + \frac{\lambda_N}{\sqrt{\beta_N}} \frac{\sqrt{I_Q} - \sqrt{i_{A1}}}{1 + \lambda_N \left(V_{TN} + \sqrt{\frac{i_{A1}}{\beta_N}} \right)} \right] i_{A1} \qquad (3.11)$$

Comparing (3.7b) and (3.11), the transfer error is reduced by a factor of 2. Hence, HD_2 and HD_3 are also reduced by a factor of 2.

3.2.3 COS Based On Cascoded Mirrors With Improved Dynamic Matching

Recently, an improved COS was proposed based on cascoded current mirrors with improved dynamic matching [13]. It is shown in Fig. 3.4. The current mirrors MA1-MA8 and MB1-MB8 provide a nominally zero transfer error.

Considering current mirror A, transistors MA1-MA3 implement a cascoded mirror with transistor MA4 acting as a common drain performing the same function as in Fig. 3.3. But a current proportional to that of the output branch is now replicated in MA5-MA7 and supplied to MA4 by means of the current mirror MA7-MA8. Unlike the circuit in Fig. 3.3, with proper design, the gate-source voltage of MA4 follows that of MA3, and, hence, the drain-source voltage of MA2 accurately matches that of MA1, even for large input currents.

Assuming the current mirror MA7-MA8 to be ideal and setting

High-drive Current Amplifiers

$$V_{SG\,A4} = V_{GS\,A3} = -V_{TP\,A4} + \sqrt{\frac{I_Q}{n\beta_{P\,A4}}} = V_{TN\,A3} + \sqrt{\frac{I_Q}{\beta_{N\,A3}}} \quad (3.12)$$

where

$$n = \frac{\beta_{A2}}{\beta_{A5}} = \frac{\beta_{A3}}{\beta_{A6}} \quad (3.13)$$

(the current I_Q/n is the bias current of MA4), MA1 and MA2 are matched in quiescent conditions. From (3.7a) the output current can be expressed as

$$i_{A2} \cong \left[1 + \lambda_N \frac{|V_{TP\,A4}| - V_{TN\,A3} + \left(\frac{1}{\sqrt{n\beta_{P\,A4}}} - \frac{1}{\sqrt{\beta_{N\,A3}}}\right)\sqrt{i_{A1}}}{1 + \lambda_N \left(V_{TN} + \sqrt{\frac{i_{A1}}{\beta_{N\,A1}}}\right)}\right] i_{A1} \quad (3.14)$$

By properly setting the aspect ratios of MA3 and MA4, the transfer error can greatly be reduced by nullifying the term in round brackets in the numerator. Thus, current i_{A2} is approximated by

$$i_{A2} \cong \left[1 + \lambda_N \left(|V_{TP\,A4}| - V_{TN\,A3}\right)\right] i_{A1} \quad (3.15)$$

Only a gain error remains which is very low if the threshold voltages are about equal.

A similar equation holds for the p-type mirror. Since i_{A2} and i_{B2} are linearly related to i_{A1} and i_{B1}, respectively, the approach discussed in Appendix 3.B has been adopted. It gives

$$HD_2 \cong \frac{2}{3\pi}\left[\lambda_N\left(|V_{TP\,A4}| - V_{TN\,A3}\right) - \lambda_P\left(V_{TN\,B4} - |V_{TP\,B3}|\right)\right] \quad (3.16a)$$

$$\cong \frac{2}{3\pi}(\lambda_N + \lambda_P)(|V_{TP}| - V_{TN})$$

$$HD_3 \cong 0 \quad (3.16b)$$

HD$_2$ is proportional to the channel-length modulation coefficients and to the differences between V_{TN} and V_{TP}.

However, (3.15) is not linear since the threshold voltage of MA3 is affected by the body effect, which was neglected when calculating harmonic distortion.

3.2.4 COS Based On Gain-Boosted Cascoded Mirrors

The last configuration to be discussed here is shown in Fig. 3.5 [13]. It uses two auxiliary differential amplifiers, A1 and A2, to force the drain voltage of the output mirror transistors to be equal to the gate voltage. Considering the n-type current mirror, we can rewrite (3.4) as

$$i_{A2} = \left(1 + \lambda_N \frac{v_{DS_{A2}} - v_{GS_{A1}}}{1 + \lambda_N v_{GS_{A1}}}\right) i_{A1} \tag{3.17}$$

Deriving (3.17) and neglecting λ^2 terms, we get

$$\frac{\partial i_{A2}}{\partial i_{A1}} \cong 1 + \lambda_N \frac{\frac{\partial v_{DS_{A2}}}{\partial i_{A1}} - \frac{\partial v_{GS_{A1}}}{\partial i_{A1}}}{1 + \lambda_N v_{GS_{A1}}} \left(\partial i_{A1} + \frac{i_{A1}}{1 + \lambda_N v_{GS_{A1}}}\right) \tag{3.18}$$

Amplifier A1 establishes the following relationship

$$\partial v_{DS_{A2}} = \frac{A}{1+A} \partial v_{GS_{A1}} - \frac{1}{1+A} \partial v_{GS_{A3}} \tag{3.19}$$

Substitution of (3.19) into (3.18) yields

$$\frac{\partial i_{A2}}{\partial i_{A1}} \cong 1 + \frac{\lambda_N}{(1+A)\sqrt{\beta_N}} \left(\frac{I_Q}{\sqrt{i_{A1}}} - 2\sqrt{i_{A1}}\right) \tag{3.20}$$

where terms $\lambda_N v_{GS_{A1}}$ and $\partial v_{GS_{A3}}/(1+A)$ have been neglected.

Following the same calculation for the p-type current mirror, and assuming the transconductance gain for both current mirrors is equal, it follows from (3.A7) and (3.A8) of Appendix 3.A that the harmonic distortion coefficients, HD_2 and HD_3, of the output current are given by

$$HD_2 \cong \frac{\lambda_N - \lambda_P}{8(1+A)\sqrt{\beta}} \left(\frac{2I_M - I_Q}{\sqrt{I_M}} \right) \qquad (3.21a)$$

$$HD_3 \cong \frac{\lambda_N + \lambda_P}{24(1+A)} \sqrt{\frac{I_Q}{\beta}} \left(2\sqrt{\frac{I_M}{I_Q}} - 1 \right) \qquad (3.21b)$$

HD_2 and HD_3 depend on the difference and the sum of λ_N and λ_P, respectively. However, the topology provides a very low-distortion COS because HD_2 and HD_3 are greatly reduced by the amplifier gain.

3.2.5 Simulation Results

To evaluate the accuracy of the proposed analysis, the circuits in Figs. 3.2-3.5 (for simplicity we shall refer to them as circuits a), b), c) and d)) were used as the output stage in the current buffer in Fig. 3.1, which was simulated using SPICE and the model parameters of a 1.2-µm CMOS process. They were biased with a current I_Q equal to 200 µA using the transistor aspect ratios in Table 3.1. In addition, real differential amplifiers (A1 and A2) were used whose gain is about equal to 30.

Table 3.1.
Transistor aspect ratios of the circuits in Figs. 3.2-3.5

Transistors	W/L (µm/µm)
MA1, MA2, MA3	600/1.4
MA4	200/1.4
MA5, MA6	60/1.4
MA7, MA8	60/3
MB1, MB2, MB3	1500/1.4
MB4	20/1.4
MB5, MB6	150/1.4
MB7, MB8	20/3

The calculated and simulated harmonic distortion of the short-circuited output current for an input signal of 100 kHz are shown in Fig. 3.6. THD decreases from circuit a) to c). Moreover, as expected, the THD of circuit b) is about half that of circuit *a*) (i.e. 6 dB better) and according to (3.15a), the THD of circuit *c*) is almost independent of the signal amplitude. In addition, HD_3 is confirmed as the main component causing distortion in circuits *a*) and *b*), and HD_2 in circuit *c*). Nevertheless, the performance of circuits *a*)-*c*) is far from what can be achieved by circuit *d*) in which THD due to channel length modulation is better than -70 dB right up to current signals which are 30 times higher than the bias current.

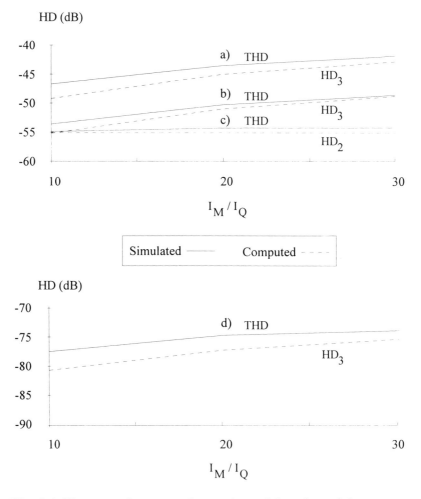

Fig. 3.6. Harmonic distortion due to channel-length modulation versus input current amplitude I_M normalized to the quiescent current I_Q, for circuits a), b) c) and d) in Fig. 3.2, 3.3, 3,4 and 3.5, respectively

For the hand calculations of HD, β_N and β_P were evaluated at about 0.012 A/V^2 and the values of the threshold voltages and channel-length modulation coefficients were extracted from the model parameters. The following approximate values were found: $\lambda_N \cong \lambda_P \cong 0.05V^{-1}$ and $V_{TN} - |V_{TP}| \cong 100mV$.

It has to be pointed out that the THD in Fig. 3.6 is much better than the typical THD performance of a real current amplifier, for two reasons. First, we use an ideal transimpedance amplifier but the finite loop gain means that non-linearity also have to be considered. Second, as is detailed below, mismatch errors in the mirroring transistors place a fundamental limit on COS linearity performance, and are the dominant sources of non-linearity in circuit d). They have not been considered in the simulation using ideally matched transistors. However, unlike the other circuits presented, in the last COS the contribution to THD due to channel-length modulation is quite negligible.

3.3 HARMONIC DISTORTION DUE TO MISMATCHES

Let us consider the effects on harmonic distortion of both threshold voltage and transconductance mismatches [9]-[11]. Mismatches almost exclusively affect the basic transistor couple of the current mirror. So, without loss of generality, we can confine our analysis to the mismatch errors in the simple current mirror MA1-MA2 in Fig. 3.1.

3.3.1 Threshold Voltage Mismatches

Neglecting the channel-length modulation and any other source of non-linearity, except that caused by mismatches in V_T, from (3.3) we can write

$$i_{A2} = \left(\frac{v_{GS\,A1,A2} - V_{TN\,A2}}{v_{GS\,A1,A2} - V_{TN\,A1}} \right)^2 i_{A1} \qquad (3.22)$$

where V_{TNA1} and V_{TNA2} are the threshold voltages of MA1 and MA2, respectively. The gate-source voltage of MA1 and MA2 is equal to

$$v_{GS\,A1,A2} = V_{TN\,A1} + \sqrt{\frac{i_{A1}}{\beta_N}} \qquad (3.23)$$

Thus, (3.22) can be written as

$$i_{A2} = \beta_N \left(\Delta V_{TN} + \sqrt{\frac{i_{A1}}{\beta_N}} \right)^2 \qquad (3.24)$$

where ΔV_{TN} is the threshold mismatch defined as

$$\Delta V_{TN} = V_{TN_{A1}} - V_{TN_{A2}} \qquad (3.25)$$

Deriving (3.24) we get

$$\frac{\partial i_{A2}}{\partial i_{A1}} = 1 + \Delta V_{TN} \sqrt{\frac{\beta_N}{i_{A1}}} \qquad (3.26)$$

By following the same steps for the *p*-type current mirror, defining its threshold mismatch $\Delta V_{TP} = V_{TP_{B1}} - V_{TP_{B2}}$, and assuming the transconductance gains to be equal (i.e. $\beta = \beta_N = \beta_P$), from (3.A7) and (3.A8) of Appendix 3.A, we obtain

$$HD_2 = \frac{1}{8} \sqrt{\frac{\beta}{I_M}} (\Delta V_{TN} - \Delta V_{TP}) \qquad (3.27a)$$

$$HD_3 = \frac{1}{24} \sqrt{\beta} \left(\frac{1}{\sqrt{I_M}} - \frac{1}{\sqrt{I_Q}} \right) (\Delta V_{TN} + \Delta V_{TP}) \qquad (3.27b)$$

In order to establish the accuracy of (3.27a) and (3.27b), SPICE simulations and the results calculated from (3.27) are shown in Figs. 3.7a and 3.7b. In evaluating the harmonic distortion components, $\Delta V_{TN} = -\Delta V_{TP}$ is chosen to estimate HD_2 and $\Delta V_{TN} = \Delta V_{TP}$ to estimate HD_3. Three curves are plotted versus the input current normalized to the bias current for different values of the percentage error $\Delta V_T/V_T$. It is thereby confirmed that HD_2 or HD_3 are the main distortion components in one of the two cases. THD is hence equal to HD_2 or HD_3 according to ΔV_{TN} and ΔV_{TP} have the same or opposite signs, respectively. Its predicted value differs from the simulated one by less than 4.5 dB. The slight difference between the calculated and simulated curves is due to the mobility degradation effect that was neglected in our analysis. In fact, surface mobility depends on the gate-source voltage, and the transconductance parameter β cannot be assumed to be constant, as has instead been done.

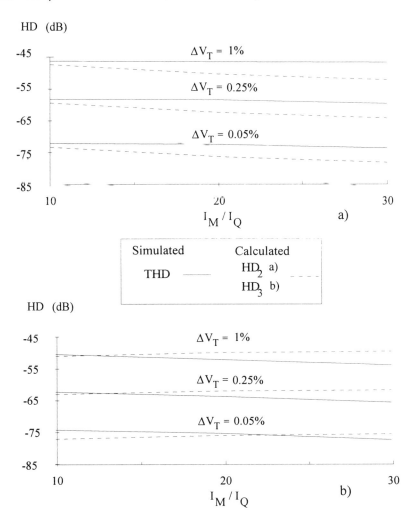

Fig. 3.7. Harmonic distortion due to threshold-voltage mismatches for a generic COS:

a) $\Delta V_{TN} = -\Delta V_{TP}$

b) $\Delta V_{TN} = \Delta V_{TP}$

3.3.2 Transconductance Parameter Mismatches

A second mismatch error which gives rise to harmonic distortion is the mismatch of the transconductance gain of the mirroring transistors. Let β_{NA1} and β_{NA2} be the transconductance gains of MA1 and MA2, and let us neglect any other source of error. From (3.4) the output current is given by

$$i_{A2} = \frac{\beta_{N_{A2}}}{\beta_{N_{A1}}} i_{A1} \tag{3.28}$$

Following the same steps for the *p*-type current mirror, from (3.B4) of Appendix 3.B it follows that

$$HD_2 \cong \frac{1}{8}\left(\frac{\beta_{N_{A2}}}{\beta_{N_{A1}}} - \frac{\beta_{P_{B2}}}{\beta_{P_{B1}}}\right) \cong \frac{1}{8}\left(\frac{\Delta\beta_N}{\beta_{N_{A1}}} - \frac{\Delta\beta_P}{\beta_{P_{B1}}}\right) \tag{3.29}$$

where $\Delta\beta_N$ and $\Delta\beta_P$ are the transconductance mismatches defined as

$$\Delta\beta_N = \beta_{N_{A2}} - \beta_{N_{A1}} \tag{3.30a}$$

$$\Delta\beta_P = \beta_{P_{B2}} - \beta_{P_{B1}} \tag{3.30b}$$

The second-order harmonic distortion is independent of the signal amplitude and is proportional to the difference between the relative transconductance mismatches. The third-order harmonic distortion is about equal to zero according to (3.A8).

Comparisons between simulations and calculations are shown in Fig. 7. Three curves are plotted for different values of the percentage error $\Delta\beta/\beta$ in the worst case condition, which is defined by setting either $\Delta\beta_N = -\Delta\beta_P$. The absence of third-order harmonic distortion is confirmed since the simulated THD is about equal to the calculated HD_2 which is underestimated by less than 0.5 dB.

From Figs. 3.7 and 3.8, the distortion caused by mismatches cannot be neglected. This is obviously true not only for the low-distortion COS *d)*, but also for circuits *a)*, *b)* and *c)* in which all the different contributions to distortion are of the same order of magnitude.

As a final remark, unlike voltage-mode amplifiers, in class AB current-mode amplifiers the dependence of overall linearity on the mismatches of the output stage does not allow such circuits to be used in applications where high linearity performance is required.

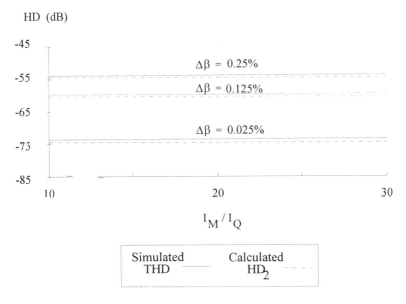

Fig. 3.8. Harmonic distortion due to transconductance-gain mismatches

3.4 DESIGN EXAMPLES

In this section we will discuss some implementations of current amplifiers based on the output stage in Fig. 3.6, which proved to be the best solution for minimizing harmonic distortion due to channel-length modulation. This COS needs some modifications for use in COAs, while it could be directly applied to VFCOAs, like the one shown in Fig. 2.27. Consequently, we shall start the discussion with a class AB VFCOA implementation.

3.4.1 A VFCOA Configuration

A simplified schematic of the current amplifier is shown in Fig. 3.9. It was presented by the authors in [24]. It is made up of two fundamental building blocks which are the transresistance amplifier and the current output stage.

The transresistance amplifier is composed of transistors M1-M12 and includes, as main blocks, a folded cascode amplifier, a common source amplifier and a class AB voltage follower. The folded cascode amplifier is made up of transistors M3-M6. The common source amplifier is made up of transistor M8 and current generator IB6. It is connected to the first stage by the common-drain M7 which provides a proper level shift. The common source amplifier drives the class AB voltage follower which is composed of transistors M9-M12. Diode-connected transistors M9 and M10 accurately set the current in M11 and M12 to a multiple (*two* in our design) of IB6.

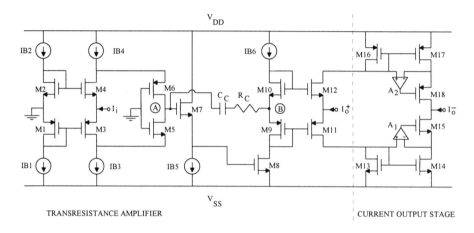

Fig. 3.9. Schematic of the current amplifier

The DC transresistance gain is given by

$$A_r(0) \cong r_{oA} g_{m8} r_{oB} \qquad (3.31)$$

where r_{oA} and r_{oB} are the output resistances of the folded cascode and common source amplifiers, respectively. The input and output resistances are given by

$$r_i \cong \frac{1}{g_{m3} + g_{m4}} \qquad (3.32)$$

$$r_{o1} = \frac{1}{g_{m11} + g_{m12}} \qquad (3.33)$$

This COS has exactly the same topology discussed in section 3.2.4 and does not require further discussion except for the implementation details relative to auxiliary amplifiers A1 and A2 and for its noise performance.

Due to the relatively low gain requirements, A1 and A2 can be implemented with a single-stage differential amplifier. Moreover, to satisfy the dynamic requirements of the upper and lower current mirrors in the COS, they are designed in complementary versions according to the schematics shown in Fig. 3.10a and 3.10b.

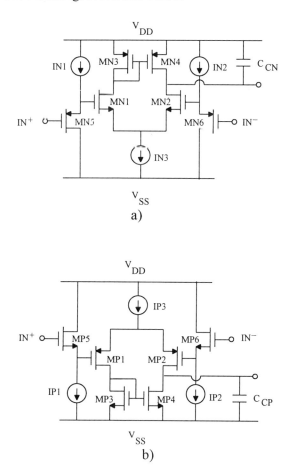

Fig. 3.10 Schematic of the auxiliary voltage amplifier A1 **a)** and A2 **b)**.

Transistors MN1-MN4 and MP1-MP4 form two differential stages, while common drain MN5-MN6 and MP5-MP6 provide a proper level shift. Dominant-pole frequency compensation is achieved by using capacitors

C_{CN} and C_{CP}. Of course, the bandwidth and slew-rate performance of such auxiliary amplifiers must be better than for the overall current amplifier to avoid speed limitations.

We have already stated that noise from a COS can make a large contribution to overall amplifier noise. Therefore, a brief noise analysis will be provided for the output stage adopted. It is easy to observe that only the basic current mirrors M13-M14 and M16-M17 contribute to the output current noise of the COS. The effect of cascode transistors, M15 and M18, is negligible since their transresistance gain is usually much smaller than one. For the same reason, the noise of auxiliary amplifiers A1 and A2 (which can be modeled by an equivalent output voltage generator) has negligible effects on the output current noise of the COS, which is given by

$$\overline{i_{nX2}^2} = 2g_{m13,14}^2 \overline{v_{n13,14}^2} + 2g_{m16,17}^2 \overline{v_{n16,17}^2} \qquad (3.34)$$

The current amplifier was compensated with the Miller capacitor C_C and resistance R_C which avoids the right half-plane zero. Assuming a unity-gain feedback configuration, the small-signal equivalent model for the loop gain of the transresistance amplifier is shown in Fig. 3.11. More specifically, g_{m8} is the transconductance gain of the common source transistor M8. C_{oB} and C_B represent the total equivalent capacitance at node B (i.e., the output of the common source stage) and that between node B and the output node, respectively. Capacitance C_i represents the equivalent capacitance due to both the output and input nodes (which are shorted in unity gain feedback).

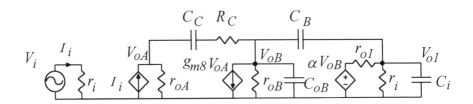

Fig. 3.11. Small-signal equivalent model of the amplifier

The loop-gain transfer function, $T(s) = -V_{o1}/V_i$, involves two zeroes and three poles whose approximated expressions are the following:

$$s_D = -\frac{1}{r_{oA}g_{m8}r_{oB}C_C} \tag{3.35a}$$

$$s_{ND1,2} = -\frac{g_{m8}C_i + \frac{1}{r_{ol}}C_{oB}}{2C_o(C_B + C_{oB})} \pm j\frac{\sqrt{4g_{m8}\frac{1}{r_{ol}}C_BC_i + 2\frac{1}{r_{ol}}\left(g_{m8}C_i - \frac{1}{r_{ol}}C_{oB}\right)C_{oB} - g_{m8}^2C_i^2}}{2C_o(C_B + C_{oB})} \tag{3.35b}$$

$$s_{Z1} = -\frac{1}{\left(R_C - \frac{1}{g_{m8}}\right)C_C} \tag{3.35c}$$

$$s_{Z2} = -\frac{1}{r_{ol}C_B} \tag{3.35d}$$

To obtain a phase margin of around 60°, after estimating C_i to be about 30 pF, C_C and R_C were set to 8 pF and 2 kΩ, respectively. Thus, the dominant-pole behavior of the loop gain is given by

$$T(s) \cong \frac{A_r(0)}{r_i + r_{ol}}\frac{1}{1 + sr_{oA}C_Cg_{m8}r_{oB}} \tag{3.36}$$

which has the following gain-bandwidth product

$$f_{GBW} \cong \frac{1}{2\pi(r_i + r_{ol})C_C} \tag{3.37}$$

In accordance with the previous equations, some guidelines can briefly be given concerning main design aspects. In respect of the transresistance amplifier, quiescent currents and transistor aspect ratios in the input stage have to take into account both noise and slew-rate requirements. In fact, high bias current reduces input noise voltage, increases slew-rate, but also increases input noise current as shown by (2.32) (2.33b). Since output noise power heavily depends on the feedback resistances (see (1.37)), making an accurate design requires knowledge of the feedback network. A trade-off between current and frequency capabilities is mandatory when designing the output stage of the transresistance amplifier. Indeed, large aspect ratios provide high signal swing but increase parasitic capacitances which affect closed-loop stability and hence the maximum achievable bandwidth.

Regarding the output stage, the large aspect ratios required for the drive capability, besides reducing frequency performance, also increase noise. However, in many cases input noise is dominant and current capability can be preserved.

Following these considerations, a current amplifier was designed. It uses transistor dimensions and bias currents for the main circuit and auxiliary differential amplifiers as reported in Table 3.2 and 3.3, respectively. The main performance of the auxiliary amplifiers is shown in Table 3.4.

Table 3.2.
Transistor aspect ratios of circuits in Fig. 3.9 and 3.10

Transistors	W/L (μm/μm)
M1, M3	160/1.4
M2, M4	80/1.4
M5	8/1.4
M6	5/1.4
M7	3/1.4
M8	90/1.4
M9	500/1.4
M10	200/1.4
M11	1000/1.4
M12	400/1.4
M13, M14, M15	600/1.4
M16, M17, M18	1500/1.4
MN1, MN2	120/2
MN3, MN4	10/4
MN5, MN6	20/2
MP1, MP2	300/2
MP3, MP4	20/4
MP5, MP6	80/2

Table 3.3.
Bias currents

Current Sources	μA
IB1, IB2, IB5	20
IB3, IB4	40
IB6	100
IN1, IN2, IP1, IP2	20
IN3, IP3	40

Table 3.4.

Main performance of auxiliary differential amplifiers A1 and A2

Parameter	A1	A2
A_d (dB)	39.6	39.6
GBW (MHz)	14	14
M_Φ (°)	61	60
V_{os} (nV)	-10.8	14.3
C_C (pF)	2	1.5

The current amplifier was fabricated in a standard 1.2-μm *n*-well CMOS process. A photo of the test chip is shown in Fig. 3.12, the total die area (except bounding pads) is 0.26 mm². The power supply was set to 5 V giving a power dissipation of 4 mW.

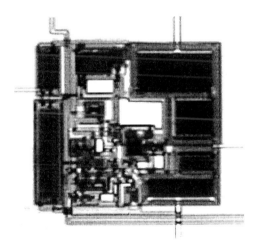

Fig. 3.12. Microphotograph of the test chip

The measured frequency response of the loop gain of the transresistance amplifier loaded with a 30-pF output capacitor is shown in Fig. 3.13. The DC gain is about 100 dB and the gain-bandwidth product is around 10 MHz with a greater than 60° phase margin. Given the high value of the DC gain, the closed-loop input and output resistances are very low. Their values in the open-loop configuration are 3 kΩ and 160 Ω, respectively. This means that the closed-loop transresistance amplifier accurately approximates an ideal transresistance amplifier, i.e. zero input and output resistances. Quite ideal output resistance is also achieved at the output of the COS thanks to the local feedback provided by A1 and A2. Its simulated value is greater than 200 MΩ.

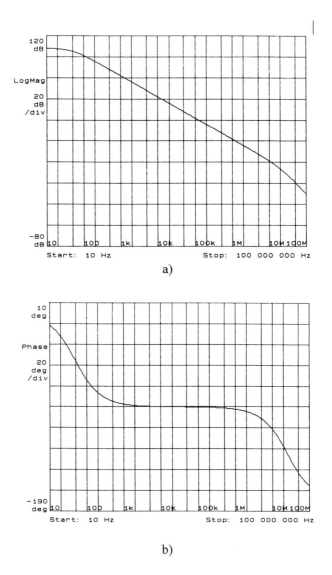

Fig. 3.13. *Loop-gain frequency response:*
a) module; b) phase

Four different frequency responses of the current amplifier in a closed-loop configuration are plotted in Fig. 3.14. The gain is varied from 0 to 30 dB only by changing resistance R_1 in Fig. 3.15.

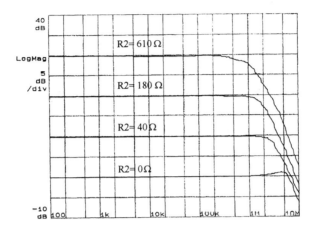

Fig. 3.14. *Closed-loop frequency responses*

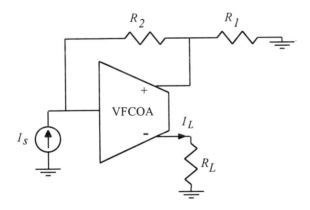

Fig. 3.15. *Closed-loop configuration*

Several measurements were carried out on the 20 available integrated samples to evaluate offset and harmonic distortion. As already discussed in sections 3.2 and 3.3, harmonic distortion in such a circuit is mainly determined by random effects on V_T and β of the transistors implementing the COS. The offset voltage was measured as being lower than 5 mV. The harmonic distortion (i.e., THD, HD_2 and HD_3) of the output current versus the input current with the amplifier in unity-gain configuration and a load resistance, R_L, of 100 Ω is plotted in Fig. 3.16.

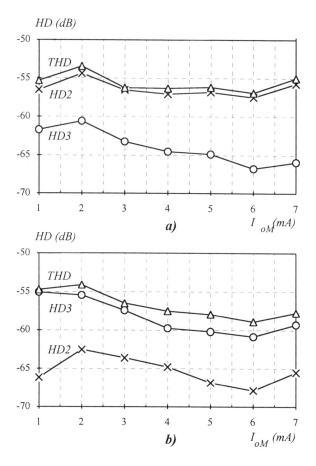

Fig. 3.16. Harmonic distortion: **a)** for the 12 samples in which HD_2 dominates; **b)** for the 8 samples in which HD_3 dominates

More specifically, two different cases were found: 12 samples in which THD is dominated by HD_2, as shown in Fig. 11a, and 8 samples in which THD is dominated by HD_3, as shown in Fig. 11b. According to (3.27a), (3.27b) and (3.29), in the samples where HD_3 is dominant, the mismatch on β is negligible with respect to that on V_T. However, for all samples the THD is almost the same and is nearly constant. Its value for an output current up to 7 mA is around -55 dB.

A 6-mA step response with the amplifier in unity-gain configuration is shown in Fig 3.17. The settling time at 0.1% is 165 ns.

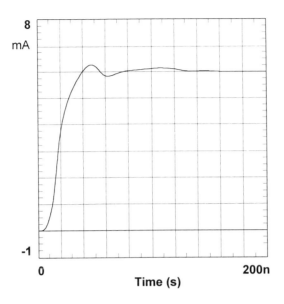

Fig. 3.17. Step response in unity-gain configuration

The power spectrum densities of the noise current and voltage generators are plotted in Fig. 3.18.

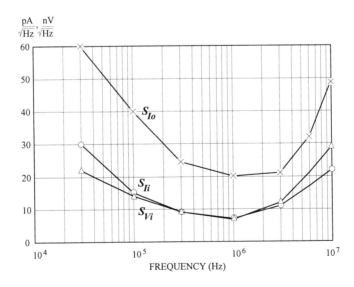

Fig. 3.18. Power spectrum densities of the noise current and voltage generators

The main electrical performance of the amplifier is summarized in Table 3.5.

Table 3.5.
Main amplifier performance

Parameter	Value
Open-Loop Gain	100 dB
GBW	10 MHz
Phase Margin	> 60°
Max. Output Current	±7 mA
Input Offset Voltage	≈ 5 mV
Output Offset Current	≈ 4.5 µA
Settling Time	165 ns
Slew Rate ($A_I = 1$)	0.2 mA/ns
THD (1 kHz, Iout = 7mA RL = 100 Ω)	< -55 dB
DC Power Dissip.	4 mW

Besides the attractive constant bandwidth feature, the solution in Fig. 3.9 is characterized by a high gain and provides a maximum bipolar current of 7 mA while dissipating 4 mW. This makes the circuit quite a good candidate as a general-purpose current amplifier. The main drawback is the absence of a differential output. Actually, the amplifier is characterized by two identical output currents. A differential output would require introducing additional current mirrors with a detrimental effect on linearity. In the next section we will show that this limitation can be avoided by adapting the same COS architecture for the implementation of a COA instead of a VFCOA.

3.4.2 COA Configurations

To provide a class AB transconductance output stage, suitable for implementing a COA, the same COS topology can be utilized. Figure 3.19 shows the schematic of a class AB current operational amplifier which uses two couples of complementary gain-boosted cascoded current mirrors in the output stage (transistors M20A-M22B and two couples of auxiliary voltage amplifiers, A1-A4) [25]. This arrangement provides a differential output stage which allows true COA operations to be achieved. Moreover, the circuit is characterized by two high impedance output nodes, thus overcoming the overdrive limitation of the traditional class AB voltage follower previously adopted.

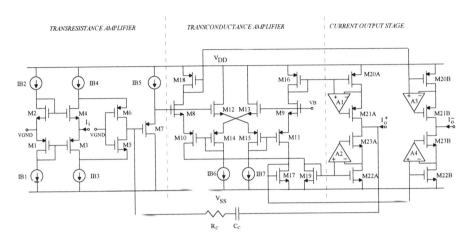

Fig. 3.19. Schematic of the class AB COA

Besides the COS, the circuit contains two other main blocks: a transresistance amplifier and a transconductance amplifier.

The transresistance amplifier is made up of transistors M1-M7 and is based on a folded cascode amplifier. It makes the main contribution to the open-loop gain of the overall amplifier. In addition, common drain M7 provides the proper bias voltage to the following stage.

The transconductance amplifier is composed of transistors M8-M19 and works in class AB fashion. It constitutes a cross-coupled differential stage which has been applied before as a nonlinear input stage in high-slew-rate op-amps [26]-[28]. In this design, its inverting input is set to the bias voltage VB which is equal to 3.5 V. In [29] a detailed analysis of the transconductance amplifier can be found. For our purposes, it is sufficient to evaluate the differential (transconductance) gain. By applying the half-circuit theory, and neglecting the common-mode signal, the simple equivalent circuit in Fig. 3.20 can be used, whose transconductance gain, g_{meq}, is given by

$$g_{meq} = \frac{i_{d8}}{v_{in}} = \frac{g_{m8} + g_{m10}}{g_{m8}g_{m10}} \qquad (3.38)$$

Of course, i_{d8} and i_{d10} are equal and properly mirrored in the COA output branches.

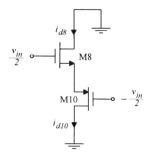

Fig. 3.20. Equivalent circuit for the gain of the transconductance stage

Frequency compensation of the main amplifier is achieved by Miller capacitor C_C and nulling resistor R_C.

As an additional comment, it can be noted that the circuit can easily be arranged in a fully-differential version, simply by adding a replica of the input stage (the transresistance amplifier) to the other input, here set to VB, of the transconductance stage.

As mentioned before, a true COA provides differential output currents. However, most applications only require a negative feedback capability which can be achieved without providing a differential output. For this purpose, a more compact solution can be implemented like the one proposed by authors [31] and shown in Fig. 3.21. The four auxiliary amplifiers of the COA in Fig. 3.19 have now been replaced by two fully differential structures, A1 and A2, which provide a balanced COS.

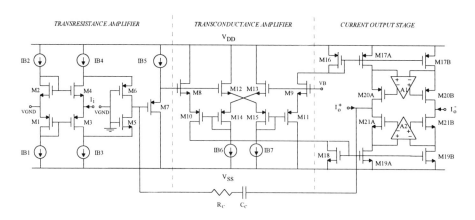

Fig. 3.21. Fabricated class AB COA

To satisfy the dynamic requirements of the upper and lower current mirror, the two auxiliary amplifiers were implemented in complementary versions according to the schematics shown in Figs. 3.22a and 3.22b.

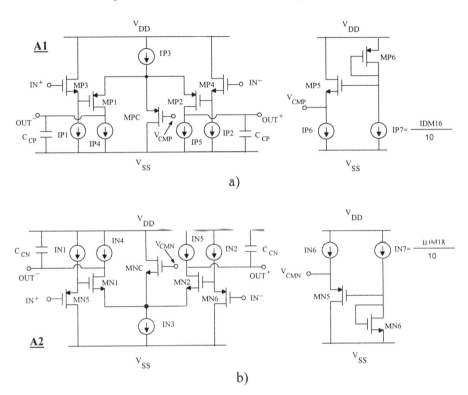

Fig. 3.22. Schematic of the auxiliary voltage amplifier A1 a) and A2 b)

The electrical components and simulated main performance of A1 and A2 are reported in Table 3.6 and 3.7. Transistors MP1-MP2 (MN1-MN2) implement the source coupled pair, while the common drains MP3-MP4 (MN3-MN4) provide a proper level shift. Dominant-pole frequency compensation is achieved by using capacitors C_{CP} and C_{CN}. Moreover, the unitary closed loop configuration allows the output common-mode voltage to be controlled very simply, by means of transistor MPC (MNC). To maximize the output swing, the common-mode voltage reference V_{CMP} (V_{CMN}) is dynamically extracted by sensing the output current using simple current mirrors.

Once again, the current amplifier has been compensated with Miller capacitor C_C and resistance R_C. Assuming a unity-gain feedback configuration, the small-signal equivalent model for the loop gain of the transresistance amplifier is shown in Fig. 3.23. Resistance r_A is the output resistance of the folded-cascode, M3-M6, and transconductance g_{meq} represents the gain of the transconductance amplifier.

Capacitance C_i is the equivalent capacitance due to both the output and input nodes (which are shorted in unity-gain feedback). CM_p and CM_n are the upper and lower current mirrors. Their nonideal frequency response is taken into account by the poles in the output current which are mainly introduced by the equivalent gate-source capacitances, C_P and C_N, associated with the current mirror itself.

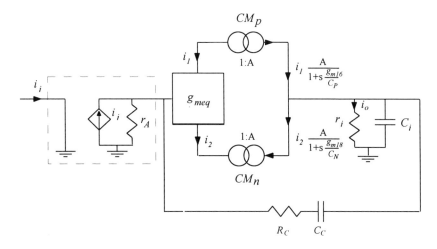

Fig. 3.23. *Small-signal equivalent model of the amplifier in Fig. 3.21*

In a first-order approximation, we can neglect the frequency limitations of the transconductance stage and consider the same time constant for the two current mirrors which means

$$i_1 = -i_2 = g_{meq} r_A i_i \qquad (3.39)$$

and

$$\frac{g_{m16}}{C_P} = \frac{g_{m18}}{C_N} = \frac{g_{mM}}{C_M} \qquad (3.40)$$

The loop-gain transfer function, $T(s) = -I_o / I_i$, is

$$T(s) = G_T r_A \frac{1 - \frac{C_C}{G_T}(1 - G_T R_C)s - \frac{C_C C_M}{g_{mM} G_T}s^2}{\left(1 + \frac{s}{G_T r_A r_i C_C}\right)\left(1 + \frac{C_i}{G_T}s + \frac{C_M C_i}{g_{mM} G_T}s^2\right)} \qquad (3.41)$$

where the total transconductance gain, G_T, is given by $G_T = 2 g_{meq} A$ and was assumed to be much higher than $1/r_i$. Equation (3.41) is useful for a pencil and paper design involving frequency stability and the gain-bandwidth product.

The dominator shows a second-order factor which is responsible for two non-dominant poles. If $\frac{g_{mM}}{C_A} > 4\frac{G_T}{C_i}$, these two poles are real and the second pole (i.e., $\frac{G_T}{C_i}$) is the one caused by the pole splitting. Unfortunately, this is not the case due to the large time constant introduced by the current mirrors in the output stage. Therefore, the poles are usually complex conjugate making frequency compensation more critical, as will become clear in the measured frequency and step response shown below.

Two real zeroes with opposite signs also appear. By setting their modules to a frequency which is much higher than *GBW*, they do not make any appreciable contribution to the loop-gain frequency response, and *GBW* can simply be given by $\frac{1}{r_i C_C}$. Of course, design optimization with SPICE will be necessary given further high-frequency poles. In our design, capacitor C_i was estimated at about 30 pF, and C_C and R_C were accordingly set to 12 pF and 200 Ω, respectively, to guarantee a phase margin of around 60°.

The power amplifier in Fig. 3.21 was fabricated in a 1.2-μm *n*-well CMOS process. Electrical parameters and transistor dimensions are shown in Tables 3.6-3.9. Table 3.10 reports the simulated main performance of the auxiliary amplifiers.

Table 3.6.
Bias currents of the main amplifier

Bias Currents	μA
IB1, IB2	20
IB3, IB4	40
IB5	20
IB6, IB7	100

Table 3.7.
Transistor aspect ratios of the main amplifier in Fig. 3.19

Transistors	W/L ($\mu m/\mu m$)
M1, M3	160/1.4
M2, M4	80/1.4
M5, M6	8/1.4
M7	10/1.2
M8, M9, M12, M13	150/1.2
M10, M11, M14, M15	350/1.2
M16	250/1.4
M17, M20	2500/1.4
M18,	100/1.4
M19, M21	1000/1.4

Table 3.8.
Transistor aspect ratios of the auxiliary amplifiers in Fig. 3.20 and 3.21

Transistors	W/L ($\mu m/\mu m$)
MP1, MP2, MPC	300/2
MP3, MP4, MP5	80/2
MP6, MP9	25/1.4
MP7, MP8	40/4
MN1, MN2, MNC	120/2
MN3, MN4, MN5	20/2
MN6, MN9	10/1.4
MN7, MN8	40/4

Table 3.9.
Bias currents of the auxiliary amplifiers

Bias Currents	μA
IP1, IP2, IP4, IP5, IP6	20
IP3	60
IN1, IN2, IN4, IN5, IN6	20
IN3	60

Table 3.10.
Simulated main performance of the auxiliary amplifiers

PARAMETERS	A1	A2
Gain (dB)	40.5	41.9
GBW (MHz)	22	20
MΦ (°)	63	68
C_C (pF)	2	2

The chip photo is shown in Fig. 3.24. The total die area is 0.63 mm^2.

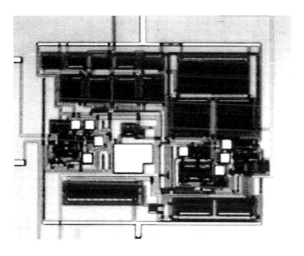

Fig. 3.24. Microphotograph of the test chip

Several measurements were carried out on experimental prototypes using a 5-V power supply which leads to a power consumption of 15 mW.

The measured frequency response of the loop gain of the amplifier loaded with a 30 pF capacitor is reported in Fig. 3.23. The DC gain, *GBW* and the phase margin are 95 dB, 8 MHz and 60°, respectively.

A 6-mA step response of the amplifier in unity-gain configuration is shown in Fig. 3.26. The settling time at 0.1 % is 260 ns while the slew rate is 0.7 mA/ns.

The complex-conjugate poles are responsible both for the slight peak in the frequency response near *GBW* and the ringing in the step response, as mentioned before.

As already shown, the linearity performance of such a circuit is mainly affected by random effects on threshold voltages and the transconductance parameters of the transistors implementing the current output stage.

Therefore, THD characterization needs statistical measurements on a large number of integrated samples. In our case, the amplifier was connected in unity-gain configuration using an output resistive load of 100 Ω. Figure 3.25 shows two worst cases in which either HD2 or HD3 dominate. In all the samples, THD is quite constant and better than -55 dB up to 14 mA. Of course, the maximum output current is limited by the voltage swing. In a short-circuit condition an output current as high as 28 mA is achieved.

The main performance of the amplifier is summarized in Table 3.11.

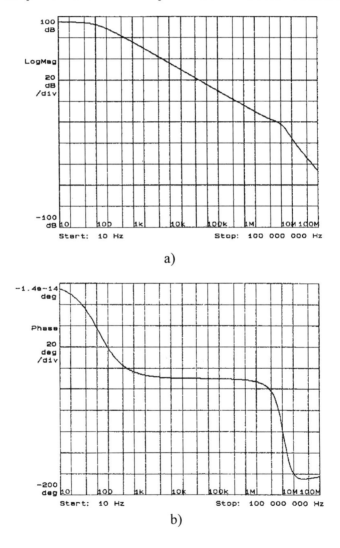

Fig. 3.25. *Frequency response of the loop gain*
a) module, b) phase

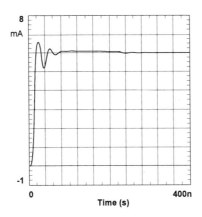

Fig. 3.26. Step response of the amplifier in unity-gain configuration

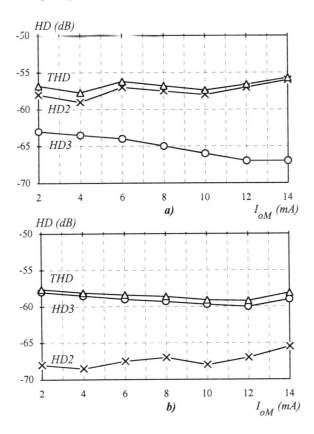

Fig. 3.27. Harmonic distortions: **a)** worst-case in which HD_2 dominates; **b)** worst-case in which HD_3 dominates.

Table 3.11.
Measured main performance

Open-Loop Gain	95 dB
GBW	8 MHz
Phase Margin	> 60°
Max. Output Current	±14 mA
Output Current Offset	< 20 µA
Settling Time (0.1%)	260 ns
Slew Rate	0.7 mA/ns
THD (1 kHz, I_{out} = 14 mA R_L = 100 Ω)	< -55 dB
DC Power Dissipation	15 mW

3.5 A VERSATILE FULLY DIFFERENTIAL COA

Following the previously mentioned idea to convert the COA in Fig. 3.19 in a fully differential version, an example of implementation is shown in Fig. 3.28 where a replica of the input stage (the transresistance amplifier) was added to the circuit in Fig. 3.19. Due to the fully differential nature of the amplifier, a circuit for the common-mode input control is required [4]. This is implemented with source followers MC1 and MC2 and the voltage to current converter composed of resistors R_1-R_2 and the common gate MC3. The common-mode reference voltage, V_{CM}, is appropriately set to maximize the output swing of the transconductance stage. Bias currents and transistor sizes of MC1-MC3 were set equal to guarantee a balanced circuit under differential-signal conditions. Considering a common-mode signal, the voltage to current converter produces a signal current which is transferred into the main circuit through the current mirrors MC4-MC4A and MC4-MC4B. A steady-state condition is achieved when voltage in nodes A and B is equal to V_{CM}. On the other hand, with differential signals no current flows into MC3 and MC4 since the current in R_1 is equal but with opposite sign of that in R_1. To preserve the overall amplifier linearity resistors R_1 and R_1 should be realized with polysilicon layers, and to avoid excessive power consumption their value should be set large. However, a large resistance value reduce both common-mode feedback gain and bandwidth, and can greatly increase silicon area.

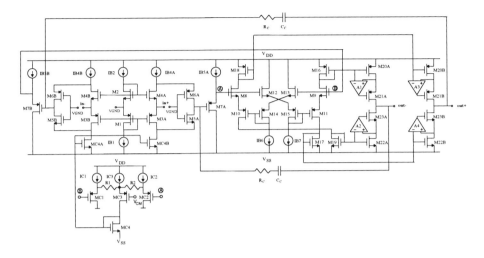

Fig. 3.28. Schematic of the fully-differential class AB COA

The circuit in Fig. 3.28 is a versatile amplifier which can be used in a large variety of applications, as illustrated below.

A. Current Amplifier

The natural application of the fully differential COA is the current amplifier shown in Fig. 3.29.

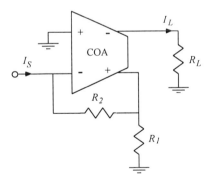

Fig. 3.29. Current amplifier block diagram

The circuit is very similar to that in Fig. 1.18. It differs from the topology previously discussed for the non-inverting input terminal which can be grounded or left open.

The closed-loop gain is given by

$$\frac{I_L}{I_S} = \left(1 + \frac{R_2}{R_1}\right) \qquad (3.42)$$

B. Transconductance Amplifier

By using an additional resistance (R_3), the current amplifier in Fig. 3.29 can easily be arranged to perform a transconductance amplifier as shown in Fig. 3.30.

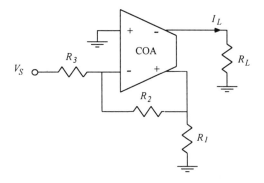

Fig. 3.30. *Transconductance amplifier block diagram*

Thanks to the virtual ground at the inverting input, resistor R_3 provides an accurate voltage to current conversion. Therefore, the closed-loop gain is given by

$$\frac{I_L}{V_S} = \frac{1}{R_3}\left(1 + \frac{R_2}{R_1}\right) \qquad (3.43)$$

B. Voltage Amplifier

The circuit in Fig. 3.28 can be configured in a fully-differential voltage amplifier, as shown in Fig. 3.31. It requires an additional common-mode feedback circuit to control the output bias voltage. This control was not used in the two previous applications (Fig. 3.29 and 3.30) since the common-mode output current is only responsible for a differential offset voltage.

The output common-mode feedback can be avoided if relative low feedback resistances (R_1 and R_2) are used. Under this condition the common-mode output current produces a small deviation from the ideal biasing condition of both the input and output terminals.

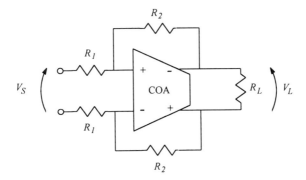

Fig. 3.31. *Voltage amplifier block diagram*

For this amplifier we get the following closed-loop gain

$$\frac{V_L}{V_S} = -\frac{R_2}{R_1} \qquad (3.43)$$

D. Transresistance Amplifier

The fully differential COA can be also arranged to implement a transresistance amplifier, as shown in Fig. 3.32.

Due to the signal current source and the floating load, no low impedance path is available for the common-mode output current. Therefore, a common-mode feedback circuit should be included at the output. However, assuming low resistive loads, an alternative approach could be used to overcome this drawback. Indeed, by using two additional resistances (R_{CM}) a low-impedance path can be provided to the common-mode current. Hence, setting R_{CM} low but greater than R_L, the common-mode resistances will not affect neither the open-loop gain nor the output swing.

The closed-loop gain is

$$\frac{V_L}{(I_1 - I_2)} = R_1 \qquad (3.44)$$

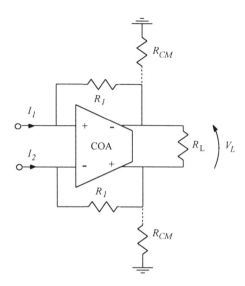

Fig. 3.32. Transresistance amplifier block diagram

3.6 MEASUREMENT STRATEGIES

As well known, a conventional measurement equipment is conceived and optimized to be used in a voltage-mode environment. Thus, to experimentally characterize circuits which accept and/or deliver currents, voltage-to-current and current-to-voltage conversions have to be provided which can simply be achieved using resistors. This approach has been extensively employed throughout the testing of the integrated samples. For instance, Fig. 3.33 illustrates the configuration adopted for the closed-loop measurements, where the *device under test* (DUT) is a current amplifier.

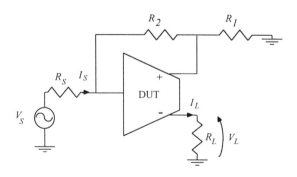

Fig. 3.33. Configuration for closed-loop measurements

Due to the virtual ground provided by the DUT, the input current I_S is equal to V_S/R_S. Therefore, the current closed-loop gain is easily expressed in terms of voltage and resistance ratios as follows

$$\frac{I_L}{I_S} = \frac{R_S}{R_L} \frac{V_L}{V_S} \qquad (3.45)$$

With this technique, characterization of gain, step response and linearity can simply be performed.

The most critical task in the experimental characterization of high-gain amplifiers is constituted by loop-gain measurements. This because the measure have to be carried out with the amplifier in open-loop configuration, but preserving operating point accuracy and loop impedance levels. To this end, the DUT was arranged in the experimental bread-board as shown in Fig. 3.34. The loop gain for the unity-gain configuration was considered.

At low frequencies, feedback is provided by resistors R_2 and R_3. They are connected between the input and the non-inverting output of the DUT. At sufficiently high frequencies capacitor C_1 is shorted and resistors R_2 and R_3 appear as input and output load, respectively. Hence, by setting R_3 and C_i to account for the input resistance and for the input and source capacitances, respectively (which were previously evaluated), the load impedance is restored at the output.

Since resistor R_1 should be very high to provide low current signals, an undesired pole could arise due to the input resistance and capacitance. To avoid this pole and to provide a further reduction of the input current, a low R_2 value can profitably be used.

At moderate frequencies C_1 is short-circuited and by inspection the circuit gives the following relationships

$$I_{in} = \frac{1}{R_1 + R_2 // r_{in}} \frac{R_2}{R_2 + r_{in}} V_s \qquad (3.46)$$

$$I_{out} = V_{out} / R_3 \qquad (3.47)$$

where r_{in} is the input resistance.

Combining (3.46) and (3.47), and considering that R_3 is equal to r_{in}, the loop gain is

$$T = \frac{I_{out}}{I_{in}} \cong \frac{R_1}{R_2} \frac{V_{out}}{V_s} \qquad (3.48)$$

The measured loop gain (i.e., V_{out}/V_s) is lower than the actual loop gain (i.e., I_{out}/I_s) by a factor R_1/R_2. This means that an accurate measurement can be carried out at high frequencies since a suitable value of the output variable (i.e., Vout) can be defined. In our board, the ratio R_1/R_2 was set to 66 dB.

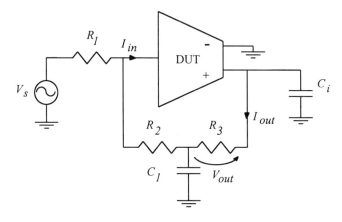

Fig. 3.34. Schematic for the loop-gain measurement

APPENDIX 3.A

Harmonic Distortion In Class AB COS

In Appendixes 3.A and 3.B, general procedures for the calculation of HD_2 and HD_3 in COS are presented, using the large-signal transfer characteristic of the current mirrors adopted.

Let us consider the output stage of the current amplifier illustrated in Fig. 3.1 and assume current i_f to be sinusoidal

$$i_f = I_M \sin(\omega t) \tag{3.A1}$$

Neglecting power terms which are higher than third-order, the output current $i_{out}(t)$ can be expressed as

$$i_{out}(t) = a_0 + a_1 i_f + a_2 i_f^2 + a_3 i_f^3 \tag{3.A2}$$

where parameter a_0 is an offset current and the gain parameter a_1 is defined as

$$a_1 \equiv \left.\frac{\partial i_{out}}{\partial i_f}\right|_0 = \left.\frac{\partial}{\partial i_f}(i_{B2} - i_{A2})\right|_0 = \left.\frac{\partial i_{B2}}{\partial i_{B1}}\right|_{I_Q} \left.\frac{\partial i_{B1}}{\partial i_f}\right|_0 - \left.\frac{\partial i_{A2}}{\partial i_{A1}}\right|_{I_Q} \left.\frac{\partial i_{A1}}{\partial i_f}\right|_0 \tag{3.A3a}$$

where I_Q is the quiescent current of the output branch.

Since we find $\Delta i_{A1} = -\Delta i_{B1} = -\frac{1}{2} i_f$, around the quiescent point, term a_1 can be written as

$$a_1 = \frac{1}{2}\left(\left.\frac{\partial i_{A2}}{\partial i_{A1}}\right|_{I_Q} + \left.\frac{\partial i_{B2}}{\partial i_{B1}}\right|_{I_Q}\right) \tag{3.A3b}$$

Thus, parameter a_1 is equal to the mirror ratio.

Parameters a_2 and a_3 can be calculated according to the approach reported in [20] and assuming the following approximation for the behavior of the class AB COS

$$i_{A1} = -i_f \quad \text{for} \quad i_f \ll -I_Q$$

$$i_{B1} = i_f \quad \text{for} \quad i_f \gg I_Q \quad (3.A4)$$

Hence, taking the derivative of i_{out} with respect to i_f, evaluated at the maximum and minimum input signal ($+I_M$ and $-I_M$), we get

$$a_2 = \frac{1}{4I_M}\left(\left.\frac{\partial_{A2}}{\partial_{A1}}\right|_{I_M} - \left.\frac{\partial_{B2}}{\partial_{B1}}\right|_{I_M}\right) \quad (3.A5)$$

$$a_3 = \frac{1}{6I_M^2}\left(\left.\frac{\partial_{A2}}{\partial_{A1}}\right|_{I_M} + \left.\frac{\partial_{B2}}{\partial_{B1}}\right|_{I_M} - 2a_1\right) =$$

$$= \frac{1}{6I_M^2}\left(\left.\frac{\partial_{A2}}{\partial_{A1}}\right|_{I_M} + \left.\frac{\partial_{B2}}{\partial_{B1}}\right|_{I_M} - \left.\frac{\partial_{A2}}{\partial_{A1}}\right|_{I_Q} - \left.\frac{\partial_{B2}}{\partial_{B1}}\right|_{I_Q}\right) \quad (3.A6)$$

Finally, HD$_2$ and HD$_3$ are given by

$$HD_2 = \frac{1}{2}\frac{a_2}{a_1}I_M \cong \frac{1}{2}a_2 I_M = \frac{1}{8}\left(\left.\frac{\partial_{A2}}{\partial_{A1}}\right|_{I_M} - \left.\frac{\partial_{B2}}{\partial_{B1}}\right|_{I_M}\right) \quad (3.A7)$$

$$HD_3 = \frac{1}{4}\frac{a_3}{a_1}I_M^2 \cong \frac{1}{4}a_3 I_M^2 = \frac{1}{24}\left(\left.\frac{\partial_{A2}}{\partial_{A1}}\right|_{I_M} + \left.\frac{\partial_{B2}}{\partial_{B1}}\right|_{I_M} - \left.\frac{\partial_{A2}}{\partial_{A1}}\right|_{I_Q} - \left.\frac{\partial_{B2}}{\partial_{B1}}\right|_{I_Q}\right)$$

$$(3.A8)$$

where a mirror ratio equal to 1 has been assumed (i.e., $a_1 = 1$).

APPENDIX 3.B

Accurate Determination Of HD_2

In class AB COSs where the two current mirrors have transfer gains which are independent of the input signal but with different values, odd harmonic distortions are zero, and even ones can be determined by applying exactly the same method described below. In fact, for these cases, this method is much more accurate than the one described in Appendix 3.A which it was found lead to an error of about 4.5 dB.

Again, consider a sinusoidal input signal with amplitude I_M, and assume an ideal class B operation with different gains a_1, b_1 for the two half waves. Currents i_{A2} and i_{B2} can be expressed as

$$i_{A2} = a_1 i_{A1} = \begin{cases} 0 & for \quad 0 < t < \frac{T}{2} \\ -a_1 i_f & for \quad \frac{T}{2} < t < T \end{cases} \quad (3.\text{B1a})$$

$$i_{B2} = b_1 i_{B1} = \begin{cases} b_1 i_f & for \quad 0 < t < \frac{T}{2} \\ 0 & for \quad \frac{T}{2} < t < T \end{cases} \quad (3.\text{B1b})$$

By expanding i_{A2} and i_{B2} in Fourier series and neglecting the even higher-order terms, we get

$$i_{A2} = -a_1 \frac{I_M}{2} \sin \omega t - \frac{2}{3\pi} a_1 I_M \cos 2\omega t \quad (3.\text{B2a})$$

$$i_{B2} = b_1 \frac{I_M}{2} \sin \omega t - \frac{2}{3\pi} b_1 I_M \cos 2\omega t \quad (3.\text{B2b})$$

Therefore, the output current is given by

$$i_{out} = i_{B2} - i_{A2} = \frac{1}{2}(a_1 + b_1)I_M \sin(\omega t) + \frac{2}{3\pi}(b_1 - a_1)I_M \cos(2\omega t)$$
(3.B3)

Assuming a mirror ratio about equal to 1, HD_2 is

$$HD_2 = \frac{2}{3\pi}(b_1 - a_1)$$
(3.B4)

REFERENCES

[1] K. Brehmer, J. Wieser, "Large Swing CMOS Power Amplifier," *IEEE J. of Solid-State Circuits*, Vol. SC-18, No. 6, pp. 624-629, Dec. 1983.

[2] R. Castello, "CMOS Buffer Amplifier," in J. Huijsing, R. van der Plassche, W. Sansen (Ed.) *Analog Circuit Design*, Kluwer Academic Publisher, 1993, pp. 113-138.

[3] P. Gray, R. Meyer, *Analysis and Design of Analog Integrated Circuit (III Ed.)*, Wiley & Sons, 1993.

[4] G. Caiulo, F. Maloberti, G. Palmisano, S. Portaluri, "Video CMOS Power Buffer with Extended Linearity," *IEEE J. of Solid-State Circuits*, Vol. 28, No. 7, pp. 845-848, July 1993.

[5] F. Eynde, W. Sansen, *Analog Interfaces for Digital Signal Processing Systems*, Kluwer Academic Publisher, 1993.

[6] P. Crawley, G. Roberts, "High-Swing MOS Current Mirror with Arbitrarily High Output Resistance," *Electronics Letters*, Vol. 28, No. 4, pp. 361-363, February 1992.

[7] E. Bruun, P. Shah, "Dynamic Range of Low-Voltage Cascode Current Mirrors," *Proc. IEEE ISCAS'95*, pp. 1328-1331, Seattle, May 1995.

[8] G. Palmisano, G. Palumbo, S. Pennisi, "High Linearity CMOS Current Output Stage," *Electronics Letters*, Vol. 31, No. 10, pp. 789-790, May 1994.

[9] J. Shyu, G. Temes, F. Krummenacher, "Random Error Effects in Matched MOS Capacitors and Current Sources," *IEEE J. of Solid-State Circuits*, Vol. SC-19, No. 6, pp. 948-955, Dec. 1984.

[10] K. Lakshmikumar, R. Hadaway, M. Copeland, "Characterization and Modeling of Mismatch in MOS Transistors for Precision Analog Design," *IEEE J. of Solid-State Circuits*, Vol. SC-21, No. 6, pp. 1057-1066, Dec. 1986.

[11] M. Pelgrom, A. Duinmaijer, A. Welbers, "Matching Properties of MOS Transistors," *IEEE J. of Solid-State Circuits*, Vol. 24, No. 5, pp. 1433-1440, Dec. 1989.

[12] G. Palmisano, G. Palumbo, S. Pennisi, "A CMOS Operational Floating Conveyor," *Proc. IEEE Midwest'94*, Lafayette, Aug. 1994.

[13] G. Palmisano, G. Palumbo, S. Pennisi, "Class AB CMOS Current Output Stages with Reduced Harmonic Distortion," *IEEE Trans. on Circuits and Systems - part II* vol. 45, no.2, pp. 243-250, Feb. 1998.

[14] K. Bult, G. Geelen, "A Fast-Settling CMOS Op Amp for SC Circuits with 90-dB DC Gain," *IEEE J. of Solid-State Circuits*, Vol. 25, No. 6, pp. 1379-1384, Dec. 1990.

[15] K. Bult, G. Geelen, "The CMOS Gain-Boosting Technique," *Int. J. Analog Integrated Circuits and Signal Processing*, No. 1, pp. 119-135, 1991.

[16] J. Lloyd, Hae-Seung Lee, "A CMOS Op Amp with Fully-Differential Gain-Enhancement," *IEEE Trans. on Circuits and Systems - part II*, Vol. 41, No. 3, pp. 241-243, March 1994.

[17] Y. Tsividis, D. Fraser, "Harmonic Distortion in Single-Channel MOS Integrated Circuits," *IEEE J. of Solid-State Circuits*, Vol. SC-16, No. 6, pp. 694-702, Dec. 1981.

[18] E. Fong, R. Zeman, "Analysis of Harmonic Distortion in Single-Channel MOS Integrated Circuits," *IEEE J. of Solid-State Circuits*, Vol. SC-17, No. 1, pp. 83-86, Feb. 1982.

[19] B. Wu, J. Mavor, "Distortion in CMOS Operational Amplifier Circuits," *IEE Proc. Part G*, Vol.131, No. 4, pp. 129-134, Aug. 1994.

[20] M. Abuelma'Atti, "Harmonic Performance of Single-Channel MOS Integrated Circuits," *IEEE J. of Solid-State Circuits*, Vol. SC-20, No. 4, pp. 860-864, Aug. 1985.

[21] M. Thoma, W. Baumann, C. Westgate, "A Method to Predict Harmonic Distorsion in Small-Geometry MOS Analog Integrated Circuits," *IEEE J. of Solid-State Circuits*, Vol. SC-22, No. 1, pp. 106-109, Feb. 1987.

[22] D. Pederson, K. Mayaram, *Analog Integrated Circuits for Communication (Principles, Simulation and Design)*, Kluwer Academic Publisher, 1991.

[23] E. Bruun, "Worst case estimate of mismatch induced distortion in complementary CMOS current mirrors," *Electronics Letters*, Vol. 34, pp. 1625-27, Aug. 1998.

[24] G. Palmisano, G. Palumbo, S. Pennisi, "High-Drive CMOS Current Amplifier", *IEEE J. of Solid-State Circuits*, Vol. 33, No.2, pp. 228-236, Feb. 1998

[25] G. Palmisano, G. Palumbo, S. Pennisi, "A High-Drive High-Gain CMOS Current Operational Amplifier", *Proc. IEEE ISCAS'98*, Monterey, May 1998.

[26] P. Li, M. Chin, P. Gray, R. Castello, "A Ratio-Independent Algorithmic Analog-to-Digital Conversion Technique," *IEEE J. of Solid-State Circuits*, Vol. SC-19, No. 6, pp. 828-836, Dec. 1984.

[27] R. Castello, P. Gray, "A High-Performance Micropower Switched-Capacitor Filter," *IEEE J. of Solid-State Circuits*, Vol. SC-20, No. 6, pp. 1122-1132, Dec. 1985.

[28] C. Wang, R. Castello, P. Gray, "A Scalable High-Performance Switched-Capacitor Filter," *IEEE J. of Solid-State Circuits*, Vol. SC-21, No. 1, pp. 57-64, Feb. 1986.

[29] E. Seevinck, R. F. Wassenaar, "A Versatile CMOS Linear Transconductor/Square-law Function Circuit," *IEEE J. of Solid-State Circuits*, Vol. SC-22, pp. 366-377, June 1987.

[30] F. Maloberti, G. Palmisano, L. Sforzini, G. Gazzoli, "Fully differential CMOS power amplifier". U.S. Patent: n° 5,281,924, Jan. 25, 1994.

[31] G. Palmisano, G. Palumbo, S. Pennisi, "A Novel CMOS Current-Mode Power Amplifier", *2nd IEEE CAS Region 8 Workshop on Analog and Mixed Design,* pp. 83-86, Baveno, Sept. 1997.

INDEX

Adjoint networks theorem, 21,3
CCII, 11,48,24-25,49-50,69,74-77,79,82
    class A, 51-63
    class AB, 63-69
CFOA, 11-16, 25-26
Channel-length modulation, 107, 110, 111, 113-115, 118, 120, 121, 125
Closed-loop amplifiers, 5
COA,
    applications, 21-23, 147-150
    definitions, 24, 3,
    implementations,74-79,136-139, 146-147
    performance parameters, 26-39
Common-mode rejection ratio, 28
Compensation
    charge injection, 91-93, 94, 96, 99, 100
    frequency, 9, 128-130, 139-141
    offset, 91
Current amplifiers, 17-21, 45-48
Current comparators,
    definitions,
    implementations,
    performance parameters,
Current gain
    differential-mode, 26
    common-mode, 27
Current input stages, 48
    class A, 51-63
    class AB, 63-69
Current output stages,
    class A, 69-74
    class AB, 108-113
Current ranges, 30
Current sensing, 18, 20
Current-mode, 1
Dynamic matching, 115-118
Dynamic range, 36
Gain-bandwidth product, 14-16, 79
Harmonic distortion, 113-124, 134, 145
Loop gain, 8, 12, 14, 37, 81, 132, 144, 152
Mismatches of
    threshold voltage, 121-123
    transconductance, 123-125

Noise,
    in current amplifiers, 32, 35-37
    in current conveyors, 50-51,53-55, 56, 59, 62, 66-68
    in current output stages, 71, 73
Offset current, 29
Offset voltage, 30
Operational amplifier, 1
Operational mirrored amplifier, 18
OTA, 10
Power supply rejection ratio, 31
Prebiasing, 86-87
Resistances, 3, 5, 32
    input, 28
    differential, 28
    common-mode, 28
Slew rate, 31
TCOA, 3, 15
Transconductance amplifier, 137-141, 148
TROA, 3, 15
VFCOA,
    definitions, 25
    implementations, 79-83
    parameters, 33, 37
VOA, 1, 3, 7, 9, 17, 22